AF556736

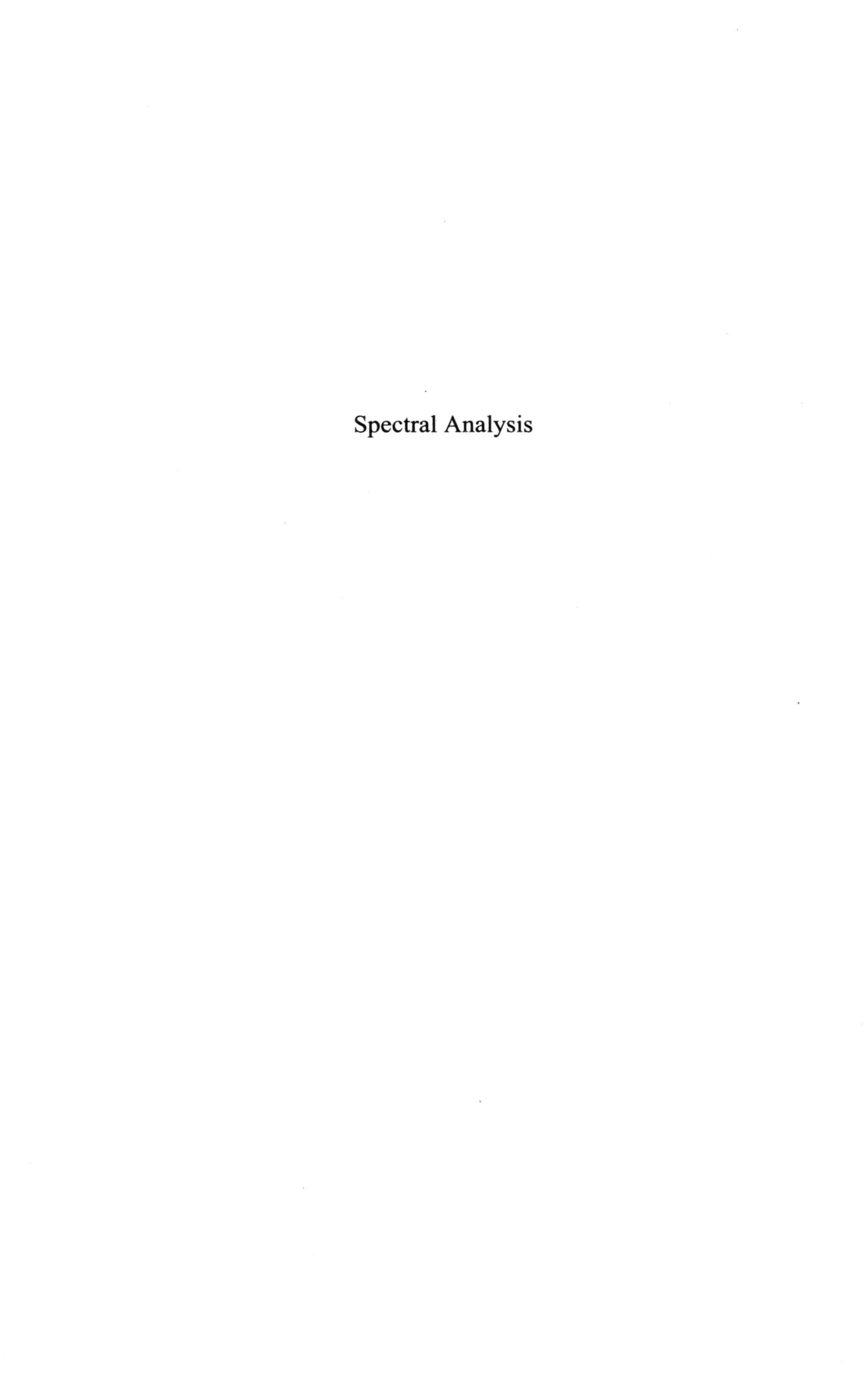

Spectral Analysis

Spectral Analysis

Parametric and Non-Parametric Digital Methods

Edited by
Francis Castanié

First published in France in 2003 by Hermes Science/Lavoisier entitled " Analyse spectrale"
First published in Great Britain and the United States in 2006 by ISTE Ltd

ISTE Ltd
6 Fitzroy Square
London W1T 5DX
UK
www.iste.co.uk

ISTE USA
4308 Patrice Road
Newport Beach, CA 92663
USA

First South Asian Edition 2007

Library of Congress Cataloging-in-Publication Data

Analyse spectrale. English
Spectral analysis: parametric and non-parametric digital methods / edited by Francis Castanié. -- 1st ed.
p. cm.
Includes bibliographical references and index.
ISBN 13: 978-1-905209-05-7
ISBN 10: 1-905209-05-3
1. Signal processing -- Digital techniques. 2. Spectrum analysis--Statistical methods. I. Castanie, Francis. II. Title
TK5102.9 .A452813
621.382'2—dc22

2006012689

British Library Cataloguing-in-Publication Data
A CIP record for this book is available from the British Library
ISBN 10: 1-905209-05-3
ISBN 13: 978-1-905209-05-7

Printed in Brijbasi Art Press Ltd., I-72, Sector-9, Noida, U.P. India.

Table of Contents

Preface

An immediate observation for those who deal with the processing of signals is the omnipresence of the concepts of frequency, which are called "spectra" here. This presence is important in the fields of theoretical research as well as in most applied sectors of engineering. We can, with good reason, wonder on the pertinence of the importance given to spectral approaches. On the fundamental plane, they relate to the Fourier transformation, projection of signals, signal descriptors, on the basis of special periodic functions, which include complex exponential functions. By generalizing this concept, the basis of functions can be wider (Hadamard, Walsh, etc.), while maintaining the essential characteristics of the Fourier basis. By projections, the spectral representations measure the "similarity" between the projected quantity and a particular base: they have henceforth no more – and no less – legitimacy than this.

The predominance of this approach in signal processing is not only based on this reasonable (but dry) mathematical description, but probably has its origins in the fact that the concept of frequency is in fact a perception through various human "sensors": the system of vision, which perceives two concepts of frequency (time-dependent for colored perception, and spatial via optical concepts of separating power), hearing, which no doubt is at the historical origin of the perceptual concept of frequency – Pythagoras and the "Music of the Spheres" – and probably other proprioceptive sensors (all those who suffer from seasickness have a direct physical experience of the frequential sensitivity).

Whatever the reasons may be, spectral descriptors are the most commonly used in signal processing; starting with this acknowledgment, the measurement of these descriptors is therefore a major issue: this is the reason for the existence of this book, dedicated to this measurement that is classically christened as *Spectral Analysis*.

It is not essential that one must devote oneself to a tedious hermeneutic to understand spectral analysis through countless books that have dealt with the subject (we will consult with interest the historical analysis of this field given in [MAR 87]). If we devote ourselves to this exercise in erudition concerning the cultural level, we realize that the theoretical approaches of current spectral analysis were structured from the late 1950s; these approaches have the specific nature of being controled by the availability of technical tools that allow the analysis to be performed. One can in this regard remember that the first spectral analyzers used optical techniques (spectrographs), then at the end of this archaeological phase – which is generally associated with the name of Isaac Newton – spectral analysis got organized around analog electronic technologies. We will not be surprised consequently that the theoretical tools were centered on concepts of selective filtering, and sustained by the theory of continuous signals (see [BEN 71]). At this time, the criteria qualifying the spectral analysis methods were formulated: frequency resolution or separating power, variance of estimators, etc. They are still in use today, and easy to assess in a typical linear filtering context.

The change to digital processing tools was first done by transposition of earlier analog transposition approaches, rewriting time first simply in terms of discrete time signals. Secondly, a simultaneous increase in the power of processing tools and algorithms opened the field up to more and more intensive digital methods. But beyond the mere availability of more comfortable digital resolution tools, the existence of such methods freed the imagination, by allowing the use of descriptors from the field of parametric modeling. This has its origin in a field of statistics known as analysis of chronological series (see [BOX 70]), which was successfully created by G. Yule (1927) for the determination of periodicities of the number of sun spots; but it is actually the availability of sufficiently powerful digital methods, in the mid 1970s, which led the community of signal processors to consider parametric modeling as a tool for spectral analysis, with its own characteristics, including the possibilities to obtain "super-resolutions" and/or to process signals of very short duration. We will see that characterizing these estimators with the same criteria as estimators from the analog word is not an easy thing: the mere quantitative assessment of the frequency resolution or spectral variances becomes a complicated problem.

The first part brings together the *processing tools* that contribute to spectral analysis. Chapter 1 lists the bases of the *signal theory* needed in order to read the following parts of the book: the informed reader could obviously skip this. Next, digital signal processing, the theory of estimation and parametric modeling of time series are presented.

The "classical" methods, known nowadays as *non-parametric methods*, form part of the second part. The privative appearing in the qualification of "non-parametric" must not be seen as a sign of belittling these methods: they are the most employed in spectral analysis.

The third and last part obviously deals with parametric methods, studying first the methods based on models of chronological series, Capron's methods and its variants, then the estimators based on the concepts of sub-spaces.

The last chapter of this part provides an opening to parametric spectral analysis of non-stationary signals, a subject with great potential, which is tackled in greater depth in another book of the IC2 series [HLA 05].

Francis CASTANIÉ

Bibliography

[BEN 71] BENDAT J. S., PIERSOL A. G., *Random Data: Analysis and Measurement Procedures,* Wiley Intersciences, 1971.

[BOX 70] BOX G., JENKINS G., *Time Series Analysis, Forecasting and Control,* Holden-Day, San Francisco, 1970.

[HLA 05] HLAWATSCH F., AUGER R., OVARLEZ J.-P., *Temps-fréquence: concepts et outils,* IC2 series, Hermès Science/Lavoisier, Paris, 2005.

[MAR 87] MARPLE S., *Digital Spectral Analysis with Applications,* Prentice Hall, Englewood Cliffs (NJ), 1987.

Specific Notations

$D(f)$ or $\text{diric}_N(t) = \dfrac{\sin(\pi N t)}{N \sin(\pi t)}$	Dirichlet's kernel
$IFT()$	Inverse Fourier transform
$DFT()$	Discrete Fourier transform
$DIFT()$	Discrete inverse Fourier transform
$\langle x(t) \rangle = \lim\limits_{T\to\infty} \dfrac{1}{T} \int_{-T/2}^{+T/2} x(t)\,dt$ or $\lim\limits_{T\to\infty} \dfrac{1}{T} \int_0^T x(t)\,dt$	Continuous time mean
$\langle x(k) \rangle = \lim\limits_{N\to\infty} \dfrac{1}{2N+1} \sum_{k=-N}^{+N} x(k)$ or $\lim\limits_{N\to\infty} \dfrac{1}{N} \sum_{k=0}^{N-1} x(k)$	Discrete time mean
$S_{xx}(f) = \hat{\gamma}_{xx}(f)$	(Auto)spectral Energy or Power density
$S_{xy}(f) = \hat{\gamma}_{xy}(f)$	Interspectral Energy or Power density
$S_{xx\ldots x}(f_1, f_2, \ldots, f_n)$	Multispectrum

PART I

Tools and Spectral Analysis

Chapter 1

Fundamentals

1.1. Classes of signals

Every signal-processing tool is designed to be adapted to one or more signal classes and presents a degraded or even deceptive performance if applied outside this group of classes. Spectral analysis too does not escape this problem, and the various tools and methods for spectral analysis will be more or less adapted, depending on the class of signals to which they are applied.

We see that the choice of classifying properties is fundamental, because the definition of classes itself will affect the design of processing tools.

Traditionally, the first classifying property is the deterministic or non-deterministic nature of the signal.

1.1.1. *Deterministic signals*

The definitions of determinism are varied, but the simplest is the one that consists of calling any signal that is reproducible in the mathematical sense of the term as a deterministic signal, i.e. any new experiment for the generation of a continuous time signal $x(t)$ (or discrete time $x(k)$) produces a mathematically identical signal. Another subtler definition, resulting from the theory of random

---

Chapter written by Francis CASTANIÉ.

signals, is based on the exactly predictive nature of $x(t)$ $t \geq t_0$ from the moment that it is known for $t < t_0$ (singular term of the Wold decomposition for example; see Chapter 4 and [LAC 00]). We will discuss here only the definition based on the reproducibility of $x(t)$, as it induces a specific strategy on the processing tools: as all information of the signal is contained in the function itself, any bijective transformation of $x(t)$ will also contain all this information. Representations may thus be imagined, which, without loss of information, will demonstrate the characteristics of the signal better than the direct representation of the function $x(t)$ itself.

The deterministic signals are usually separated into classes, representing integral properties of $x(t)$, strongly linked to some quantities known by physicists.

Finite energy signals verify the ıntegral properties [1.1] and [1.2] with continuous or discrete time

$$E = \int_R |x(t)|^2 dt < \infty \tag{1.1}$$

$$E = \sum_{k=-\infty}^{+\infty} |x(k)|^2 < \infty \tag{1.2}$$

We recognize the membership of $x(t)$ to standard function spaces (noted as L_2 or l_2 respectively), as well as the fact that this integral, to within some dimensional constant (an impedance in general), represents the energy E of the signal.

Signals of *finite average power* verify:

$$P = \lim_{T \to \infty} \frac{1}{T} \int_{-T/2}^{+T/} |x(t)|^2 dt < \infty \tag{1.3}$$

$$P = \lim_{N \to \infty} \frac{1}{2N+1} \sum_{k=-N}^{+N} |x(k)|^2 < \infty \tag{1.4}$$

If we accept the idea that the sums of equation [1.1] or [1.2] represent "energies", those of equation [1.3] or [1.4] then represent powers.

It is clear that these integral properties correspond to mathematical characteristics whose morphological behavior along the time axis is very different: the finite energy signals will be in practice "pulse shaped", or "transient" signals such that $|x(t)| \to 0$ for $|t| \to \infty$. This asymptotic behavior is not at all necessary to ensure the convergence of the sums, and yet all practical finite energy signals verify it. As a simple example, the signal below is of finite energy:

$$x(t) = Ae^{-at}\cos(2\pi ft + \theta) \qquad t \geq 0$$
$$= 0 \qquad t < 0$$

(this type of damped exponential oscillatory waveform is a fundamental signal in the analysis of linear systems that are invariant by translation).

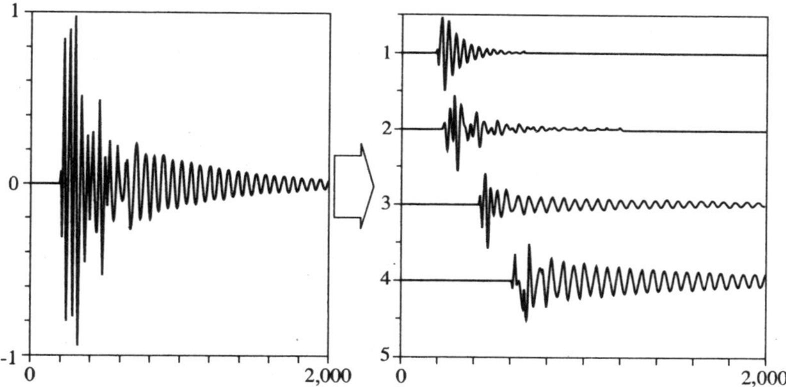

Figure 1.1. *Electromagnetic interference signal, and its decomposition*

On the more complex example of Figure 1.1, we see a finite energy signal of the form:

$$x(t) = \sum_{i=1}^{4} x_i(t - t_i)$$

where the 4 components $x_i(t)$ start at staggered times $\{t_i\}$. Its shape, even though complex, is supposed to reflect a perfectly reproducible physical experiment.

Finite power signals will be, in practice, *permanent* signals, i.e. not canceling at infinity. As a simple example:

$$x(t) = A(t)\sin(\psi(t)) \quad t \in \Re$$
$$|A(t)| < \infty \qquad [1.5]$$

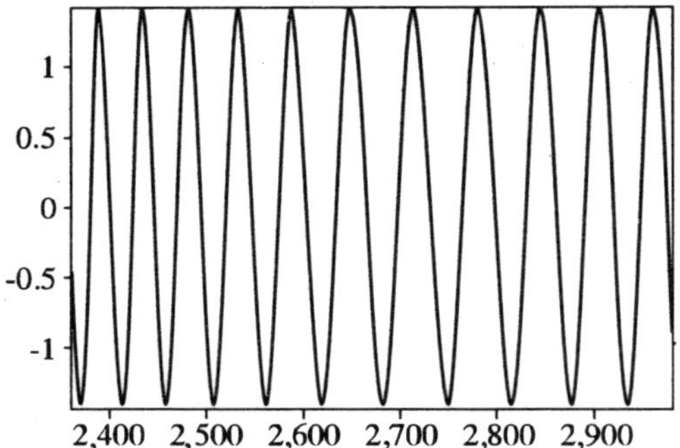

Figure 1.2. *Frequency modulated sine curve*

An example is given in Figure 1.2. It is clear that there is obviously a start and an end, but its mathematical model cannot take this unknown data into account, and it is relevant to represent it by an equation of the type [1.5]. This type of signal, modulated in amplitude and angle, is fundamental in telecommunications.

1.1.2. *Random signals*

The deterministic models of signals described in the previous paragraph are absolutely unrealistic, in the sense that they do not take into account any inaccuracy, or irreproducibility, even if partial. To model the uncertainty on the signals, several approaches are possible today, but the one that was historically adopted at the origin of the signal theory is a probabilistic approach.

In this approach, the signal is regarded as resulting from a drawing in a probability space. The observation $x_j(t)$ (j^{th} outcome, or "realization") is one of the results of the random event, but is no longer *the* signal: the latter is now constituted by all the possible outcomes $x(t, \omega)$. The information of the signal is no longer contained in one of the realizations $x_j(t)$, but in the probability laws that govern $x(t, \omega)$. Understanding this fact is essential: it is not relevant, in this approach, to look for representations of the signal in the form of transformations on one of the outcomes that could well demonstrate the properties sought; here we must "go back" as far as possible to the probability laws of $x(t, \omega)$, from the observation of one (or a finite number) of the outcomes, generally over a finite time interval. How, from the data $\{x_j(t)\, j = 1,\ldots,K, t \in [t_0, t_1]\}$, we can go back to the laws of $x(t, \omega)$ is the subject of the estimation theory (see Chapter 3).

The example of Figure 1.3 shows some outcomes of a same physical signal (Doppler echo signal from a 38 GHz traffic radar carried by a motor car). Every realization contains the same relevant information: same vehicle speed, same pavement surface, same trajectory, etc. However, there is no reproducibility of outcomes, and there are fairly large morphological differences between them. At the very most we cannot avoid noticing that they have a "family likeness". The probabilistic description is nothing but one of the approaches that helps account for this "family likeness", i.e. by the membership of the signal to a well-defined probability space.

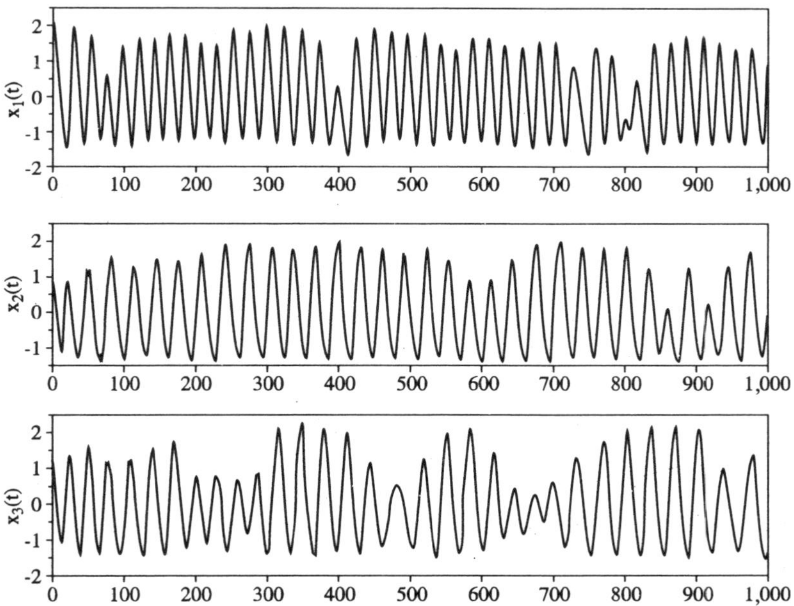

Figure 1.3. *Example of 3 outcomes of an actual random signal*

The laws are probability sets

$$P\left[x(t_1)\in D_1, x(t_2)\in D_2,\ldots,x(t_n)\in D_n\right] = L(D_1,D_2,\ldots,D_n,t_1,t_2,\ldots,t_n) \quad [1.6]$$

whose most well-known examples are the probability densities in which $D_i = [x_i, x_i + dx_i[$

$$P[\ldots] = p(x_1,\ldots,x_n,t_1,\ldots,t_n)\,dx_1,\ldots,dx_n \quad [1.7]$$

and the distribution functions, with $D_i =]-\infty, x_i]$

$$P[\ldots] = F(x_1,\ldots,x_n,t_1,\ldots,t_n) \quad [1.8]$$

The set of these three laws fully characterizes the random signal, but a partial characterization can be obtained *via* the moments of order M of the signal, defined (when these exist) by:

$$\begin{aligned} E\left(x(t_1)^{k_1} \,..x(t_n)^{k_n}\right) &= \int\cdots\int_{\Re^n} x_1^{k_1}\ldots x_n^{k_n}\, p(x_1,\ldots,x_n,t_1,\ldots,t_n)\, dx_1,\ldots,dx_n \\ &= \int\cdots\int_{\Re^n} x_1^{k_1}\ldots x_n^{k_n}\, dF(x_1,\ldots,x_n,t_1,\ldots,t_n) \quad [1.9] \\ \text{with } M &= \sum_i k_i \end{aligned}$$

It can be shown that these moments are linked in a simple way to the Taylor series expansion of the characteristic function of the n-tuple $\{x_1(t_1),\ldots,x_n(t_n)\}$ defined by:

$$\begin{aligned} &\phi(u_1,\ldots,u_n) = \phi(\mathbf{u}) \triangleq E\left(\exp\left(j\mathbf{u}^T.\mathbf{x}\right)\right) \\ &\mathbf{u} = |u_1 \,\ldots\, u_n|^T,\ \mathbf{x} = |x_1 \,\ldots\, x_n|^T \end{aligned} \quad [1.10]$$

We see that the laws and the moments have a dependence on the considered points $\{ti,\ldots,t_n\}$ of the time axis. The separation of random signals into classes very often refers to the nature of this dependence: if a signal has one of its characteristics invariant by translation, or in other words, independent of the time origin, we will call the signal stationary for this particular characteristic. We will thus speak of *stationarity in law* if:

$$p(x_1,\ldots,x_n,t_1,\ldots,t_n) = p(x_1,\ldots,x_n,t_1+t_0,\ldots,t_n+t_0)$$

and of *stationarity for the moment of order M* if:

$$E\left(x(t_1)^{k_1}\ldots x(t_n)^{k_n}\right) = E\left(x(t_1+t_0)^{k_1} \,..x(t_n+t_0)^{k_n}\right)$$

The interval t_0 for which this type of equality is verified holds various stationarity classes:

– *local* stationarity if this is true only for $t_0 \in [T_0, T_1]$,

– a*symptotic* stationarity if this is verified only when $t_0 \to \infty$, etc.

We will refer to [LAC 00] for a detailed study of these properties.

Another type of time dependence is defined when the considered characteristics recur *periodically* on the time axis: we then speak of *cyclostationarity*. For example, if:

$$E\left(x(t_1)^{k_1}\ldots x(t_n)^{k_n}\right)=E\left(x(t_1+T)^{k_1}\ ..x(t_n+T)^{k_n}\right)\forall(t_1,\ldots,t_n)$$

the signal will be called cyclostationary for the moment of order M (see for example [GAR 89]).

1.2. Representations of signals

The various classes of signals defined above can be represented in many ways, i.e. forming of the information present in the signal, which does not have any initial motive other than to highlight certain characteristics of the signal. These representations may be complete, if they carry all the information of the signal, and partial otherwise.

1.2.1. *Representations of deterministic signals*

1.2.1.1. *Complete representations*

As the postulate of the deterministic signal is that any information that affects it is contained in the function $x(t)$ itself, it is clear that any bijective transformation applied to $x(t)$ will preserve this information.

From a general viewpoint, such a transformation may be written with the help of integral transformations on the signal, such as:

$$R(u_1,\ldots u_m)=H\left[\int\cdots\int G\left[x(t_1),\ldots,x(t_n)\right]\cdot\psi(t_1,\ldots,t_n,u_1,\ldots,u_m)\ldots dt_1\ldots dt_n\right]\quad[1.11]$$

where the operators H [.] and G [.] are not necessarily linear. The kernel $\psi(t,f)=e^{-j2\pi ft}$ has a fundamental role to play in the characteristics that we wish to display. This transformation will not necessarily retain the dimensions, and in general we will use representations such as $m\geq n$. We will expect it to be inversible (i.e. there exists an exact inverse transformation), and that it helps highlight certain properties of the signal better than the function $x(t)$ itself. It must also be understood that the demonstration of a characteristic of the signal may considerably mask other characteristics: the choice of H [.], G [.] and $\psi(.)$ governs the choice of what will be highlighted in $x(t)$. In the case of discrete time signals, discrete expressions of the same general form exist.

The very general formula of equation [1.11] fortunately reduces to very simple elements in most cases.

Here are some of the most widespread examples: orthogonal transformations, linear time-frequency representations, time scale representations, and quadratic time-frequency representations.

Orthogonal transformations

This is the special case where m = n = 1, $H[x] = G[x] = x$ and $\psi(t, u)$ is a set of orthogonal functions. The undethronable Fourier transform is the most well-known with $\psi(t,f) = e^{-j2\pi ft}$ We may once again recall its definition here:

$$\tilde{x}(f) = FT(x(t)) = \int_{\Re} x(t)e^{-j2\pi ft}dt \text{ with } f \in \Re \qquad [1.12]$$

and its discrete version for a signal vector $\mathbf{x} = |x(0)\ldots x(N-1)|^T$ of length N:

$$\tilde{x}(l) = DFT(x(k)) = \sum_{k=0}^{N-1} x(k)e^{-j2\pi\frac{kl}{N}} \text{ with } l \in [0, N-1] \qquad [1.13]$$

If we bear in mind that the inner product appearing in equation [1.12] or [1.13] is a measure of likelihood between $x(t)$ and $e^{-j2\pi ft}$, we immediately see that the choice of this function set $\psi(t,f)$ is not innocent: we are looking to highlight complex sine curves in $x(t)$.

If we are looking to demonstrate the presence of other components in the signal, we will select another set $\psi(t,u)$. The most well-known examples are those where we look to find components with binary or ternary values. We will then use families such as Hadamard-Walsh or Haar (see, for example, [AHM 75]).

Linear time-frequency representations

This is the special case where $H[x] = G[x] = x$ and $\psi(t,\tau,f) = w(t-\tau)e^{-j2\pi ft}$ leading to an expression of the type:

$$R(\tau,t) = \int x(t)w(t-\tau)e^{-j2\pi ft}dt \qquad [1.14]$$

Time scale (or wavelet) representations

This is the special case where $H[x] = G[x] = x$ and $\psi(t,\tau,a) = \psi_M^*\left(\frac{t-\tau}{a}\right)$ which gives:

$$R(\tau,a) = \int x(t)\, \psi_M^*\left(\frac{t-\tau}{a}\right) dt \qquad [1.15]$$

where $\psi_M(t)$ is the parent wavelet. The aim of this representation is to look for the parent wavelet in $x(t)$, translated and expanded by the factors τ and a respectively.

Quadratic time-frequency representations

In this case, G [.] is a quadratic operator, and H [.] = x. The most important example is the Wigner-Ville transformation:

$$R(t,f) = \int x\left(t+\frac{\tau}{2}\right) x\left(t-\frac{\tau}{2}\right)^* e^{-j2\pi f\tau} d\tau \qquad [1.16]$$

which is inversible if at least one point of $x(t)$ is known.

The detailed study of these three types of representations is the central theme of another book of the same series (see [HLA 05]).

Representations with various more complex combinations of H [x], G [x], and ψ_M (.) can be found in [OPP 75], [NIK 93], from among abundant literature.

1.2.1.2. *Partial representations*

In many applications, we are not interested in the entire information present in signal, but only in a part of it. The most common example is that of energy representations, where only the quantities linked to the energy or power are considered.

The energy or power spectral density

If we are interested in the frequency energy distribution (we will call it "spectral"), we are led to consider the following quantity:

$$S_x(f) = \left|FT(x(t))\right|^2 = \left|\hat{x}(f)\right|^2 \triangleq ESD(x(t)) \qquad [1.17]$$

for continuous time signals, and:

$$S_x(l) = \left|DFT(x(k))\right|^2 = \left|\hat{x}(l)\right|^2 \triangleq ESD(x(k)) \qquad [1.18]$$

for discrete time signals.

These representations do not allow the signal to be reconstructed, as the phase of the transform is lost. We will call them ESD (energy spectral densities). But the main reason for interest in these partial representations is that they show the energy distribution along a non-time axis (here the frequency). In fact, Parseval's theorem states:

$$E = \int_R |x(t)|^2 dt = \int_R |\hat{x}(f)|^2 df \qquad [1.19]$$

and $|\hat{x}(f)|^2$ represents the energy distribution in f, just as $|x(t)|^2$ therefore represents its distribution over t.

For discrete time signals, Parseval's theorem is expressed a little differently:

$$E = \sum_{k=0}^{N-1} |x(k)|^2 = \frac{1}{N}\sum_{l=0}^{N-1} |\hat{x}(l)|^2$$

but the interpretation is the same.

The quantities $S_x(f)$ or $S_x(l)$ defined above have an inverse Fourier transform, which can easily be shown as being expressed by:

$$IFT\left(S_x(f)\right) = \int_R x(t)\, x^*(t-\tau)\, dt \triangleq \gamma_{xx}(\tau) \qquad [1.20]$$

This function is known as the autocorrellation function of $x(t)$. We see that the ESD of $x(t)$ could have axiomatically been defined as follows:

$$S_x(f) = ESD\left(x(t)\right) \triangleq FT\left(\gamma_{xx}(\tau)\right) \qquad [1.21]$$

When we are interested in *finite power* signals, their Fourier transform exists in terms of distributions. This generally prohibits the definition of the square of the modulus, and thus the PSD as by equation [1.17] or [1.18]. It is, however, always possible to define the autocorrelation function by:

$$\gamma_{xx}(\tau) \triangleq \lim_{T\to\infty} \frac{1}{T} \int_{-T/2}^{+T/2} x(t)\, x^*(t-\tau)\, dt \qquad [1.22]$$

Its Fourier transform is used to define the PSD by:

$$S_x(f) = PSD\left(x(t)\right) \triangleq FT\left(\gamma_{xx}(\tau)\right) \qquad [1.23]$$

This function gives the frequency distribution of the power, such that its sum along the axis f once again gives the total power:

$$\int_{\Re} S_x(f)\,df = \gamma_{xx}(0) = P \qquad [1.24]$$

All this is greatly detailed, for example in [DUV 91], and taken up in section 2.2 of this book.

The reader will easily understand that the previously mentioned definitions of energy (or power) spectral densities are based on the Fourier transform, only for the isometry represented by equation [1.19]. We can thus easily generalize these concepts through other orthogonal transforms (Hadamard-Walsh, Haar, etc.) that have the same property of preserving the inner product. The axis along which the ESD will give the energy distribution will no longer be a frequency axis, but an axis corresponding to the subscripting of the family of orthogonal functions retained (see [HAR 69], for example).

1.2.2. *Representations of random signals*

1.2.2.1. *General approach*

As specified in section 1.1.2, the complete knowledge of the information present in the signal requires the knowledge of the probability laws. This occurs in practice only when we have a theoretical model establishing the law, and only its parameters remain to be determined: this is a problem that therefore deals with the theory of parametric estimation (see Chapter 3).

Except for this slightly ideal case, the representations of random signals will only be partial. Among these representations, the moments and cumulants are the most used (if not the only ones). We will talk of *knowledge of the order K* of the signal if all moments up to the K order are known, i.e. the set of:

$$E\left(x(t_1)^{k_1}\ldots x(t_n)^{k_n}\right)\text{for } M = 1,2,\ldots,K$$

$$\text{with } M = \sum_i k_i$$

Each moment is a function of n variables, and it is obvious that for a high value of K, the representation will be very complex. The practical ambitions are largely limited, and we practically limit ourselves to $K = 2$, and in a few cases to $K = 3$ or 4. For a definition and presentation of cumulants, refer to [LAC 97] and [NIK 93].

1.2.2.2. *2nd order representations*

The most important case is that of the 2nd order representation, which reduces to:

$$\begin{aligned} m_x(t) &= E\big(x(t)\big) \\ \gamma_{xx}(t_1,t_2) &= E\big(x(t_1)x^*(t_2)\big) \end{aligned} \qquad [1.25]$$

We see that this representation requires the use of a single variable (the average), and a function with 2 variables (the autocorrelation function).

In the case where we can, additionally, make the stationary hypothesis, this reduces to a constant and a function with 1 variable:

$$\begin{aligned} m_x &= E\big(x(t)\big) \\ \gamma_{xx}(\tau) &= E\big(x(t)x^*(t-\tau)\big) \end{aligned} \qquad [1.26]$$

Considering the complexity of representations as the order increases, we can presume that a limitation to such low orders is only for the sake of simplicity: it would therefore be troublesome if the sacrifice made for the sake of simplicity is too great, and that the essence of the signal information is lost in this limitation.

There exists a very important class of random processes for which the knowledge of the 2nd order is complete: this is the class of Gaussian signals, for which all laws are fully determined, for all orders, if one knows the quantities given by equation [1.25] (or [1.26]) *stricto sensu* for all $(t,t_1,t_2)\in\Re^3$. We will refer to [LAC 00] for a detailed representation of these signals. But for the essence, their abundance among actual signals results from asymptotic theorems on linear mixtures of laws that are (almost) arbitrary, which have an inevitable attraction for the Gaussian law.

In addition, other classes of signals enjoy the same parsimony of complete representations: certain non-linear transformations of Gaussian signals (such as those found in telecommunications), compound random processes, invariant spherical processes, etc.

Even if the 2nd order representation remains partial, it is of great interest as it is consistent with an "energy" representation of random signals, very close to the one described in section 1.2.1.2.

In fact, the ergodic theorem (see [DUV 91, LAC 00], for example) tells us that, in the field of application of this theorem, there is an identity between the mathematical expectation operator E [.] and that of the time average $\lim_{T\to\infty} \frac{1}{T} \int_{-T/2}^{+T/2} .dt$ applied to any characterization $x_j(t)$. For example, the 1st order ergodicity can be used to write:

$$E\left[x(t)\right] = \lim_{T\to\infty} \frac{1}{T} \int_{-T/2}^{+T/2} x_j(t)\,dt \triangleq \left\langle x_j(t)\right\rangle \forall j \tag{1.27}$$

The stochastic average and the time average – which is known as the *direct current* (DC) for electrical quantities – are therefore, in this case, confounded.

The 2nd order ergodicity is highly revealing; in fact:

$$E\left[\left|x(t)\right|^2\right] = \lim_{T\to\infty} \frac{1}{T} \int_{-T/2}^{+T/2} \left|x_j(t)\right|^2 dt = \left\langle \left|x_j(t)\right|^2 \right\rangle = P \tag{1.28}$$

In the time average appearing in the second member, we recognize the average power of the jth realization (which is independent of j in this case). The Bochner-Khintchine theorem allows us to make the link between the 2nd order moment (autocorrelation function) and the Power Spectral Density (PSD) (in terms of stationarity random signals):

$$\begin{aligned} &\gamma_{xx}(\tau) \triangleq E\left(x(t)\,x^*(t-\tau)\right) \\ &S_{xx}(f) \triangleq FT\left(\gamma_{xx}(\tau)\right) \end{aligned} \tag{1.29}$$

and from there, we find the form similar to equation [1.23]

$$\begin{aligned} &\gamma_{xx}(0) = \int_{\Re} S_x(f)\,df = P \\ &\gamma_{xx}(0) = E\left[\left|x(t)\right|^2\right] = \lim_{T\to\infty} \frac{1}{T} \int_{-T/2}^{+T/2} \left|x_j(t)\right|^2 dt = \left\langle \left|x_j(t)\right|^2 \right\rangle = P \end{aligned}$$

The power figuring in the right-hand term of this equation corresponds to the usual definition of power using a time average. The PSD as defined for the stationarity and ergodic random processes is therefore identical to the PSD as defined by finite power signals, applied to any realization of the signal.

One must, however, note that this similarity does not apply as soon as we leave the stationary and ergodic framework: the loss of one of the two properties prohibits the two definitions of the PSD, with only equation [1.29] remaining valid. We will meet this type of divergence in Chapter 9.

Very complete theoretical results were established for the spectral density classes. This set of properties results essentially from the fact that the *class of correlation functions* and that of *characteristic functions* (from probability calculus) are confounded (differing by a multiplicative constant). This very significant result helps take advantage of all historical work on characteristic functions. One of the essential consequences is that the nature of this mathematical object is well worked out; in particular, a very general decomposition helps affirm that for any 2nd order stationarity signal [LAC 00]:

$$S_{xx}(f) = S_1(f) + S_2(f) + S_3(f) \qquad [1.30]$$

where the 3 components have well established properties:

The component $S_1(f)$ is the "continuous" part of the PSD. This name is not strictly correct: it is only the derivative of a continuous increasing function in the wide sense. It can thus have discontinuities, asymptotes, etc., but on a set of zero measure. The summability of $S_1(f)$ generally ensures "pleasant" behavior for $|f| \to \infty$ that is to say $S_1(\infty) = 0$ (although this is not a necessary condition).

The component $S_2(f)$ is the derivative of a non-decreasing step function. Its compact expression is therefore:

$$S_2(f) = \sum A_i \delta(f - f_i) \qquad [1.31]$$

where each $A_i \delta(f - f_i)$ is a monochromatic component (known as "line") of power $A_i \geq 0$ and frequency f_i. The summability of $S_2(f)$ therefore ensures that of the series $\{A_i\}$, i.e. $P_2 = \sum_i A_i < \infty$. The term P_2 represents the total power of periodic components of the signal. We say that $S_2(f)$ is a *line spectrum.*

The special case where a sub-set of the series $\{f_i\}$ consists of rational numbers between mutually rational values corresponds to a *periodic component* in the random signal. This type of component deserves a special development, as it is not intuitively evident that a signal can be random and stationarity as well as periodic at the same time.

However, if we define the periodicity as in the mean square sense of a signal $x(t)$ with:

$$E\left(\left|x(t) - x(t + mT)\right|^2\right) = 0 \ \forall m \in \mathbb{Z} \qquad [1.32]$$

With T representing the period of the random signal $x(t)$, this can be broken down into a Fourier series with *random coefficients*:

$$C_k = \frac{1}{T} \int_{-T/2}^{+T/2} x(t) e^{-j2\pi k \frac{t}{T}} dt \qquad [1.33]$$

The random variables $\{C_k\}$ have the following striking properties:

$$\begin{aligned} &E(C_k) = 0 \text{ for } k \neq 0 \\ &E\left(C_k C_m^*\right) = 0 \text{ for } k \neq m \\ &E\left(|C_k|^2\right) = A_k \end{aligned} \qquad [1.34]$$

where the power A_k is associated with the line at frequency $f_k = k/T$ appearing in equation [1.31].

Note that there exists a line for $f = 0$ (by convention, we note it as $f_0 = 0$) which is linked to what is commonly known as the "DC component", as it is easy to show that

$$A_0 = \left|E(x(t))\right|^2 \qquad [1.35]$$

where the average $E(x(t))$ is identical to the continuous component of electronics engineers, if the signal is ergodic (see equation [1.27]).

The term $S_3(f)$ is known as singular and is, generally, absent in physical signals.

The previous properties are very important to understand what kind of object we will be confronted with when we will perform a 2nd order spectral analysis on stationary signals. Figure 1.4 summarizes the properties of these PSD.

The main role of the autocorrelation function and its Fourier transform, the PSD, is largely justified by the regularity of their behavior in the invariant linear transformations. We must briefly recall here that if two random signals are linked by an invariant linear filtering operation of kernel $h(t)$ (continuous time) or $h(k)$ (discrete time), a kernel that is also its impulse response, we can write it in the convoluted form:

$$\begin{aligned} &y(t) = x \otimes h(t) = \int_{\Re} x(u) h(t-u) du \\ &y(k) = x \otimes h(k) = \sum_{m=-\infty}^{+\infty} x(m) h(k-m) \end{aligned} \qquad [1.36]$$

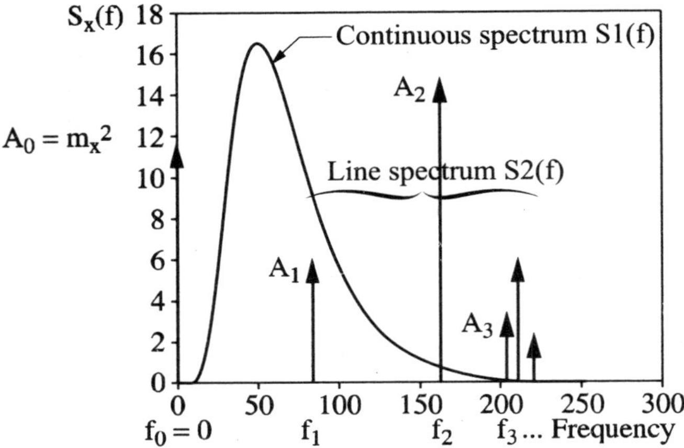

Figure 1.4. *Morphology of PSDs in the stationarity case*

The Wiener-Lee (or Blanc-Lapierre) theorems can be used to write:

$$\begin{aligned} S_{yx}(f) &= S_x(f).\hat{h}(f) \underset{\longleftarrow}{FT} \gamma_{yx}(\tau) = \gamma_{xx} \otimes h(\tau) \\ S_{yy}(f) &= S_{xx}(f)\left|\hat{h}(f)\right|^2 \underset{\longleftarrow}{FT} \gamma_{yy}(\tau) = \gamma_{xx} \otimes h \otimes h^-(\tau) \end{aligned} \quad [1.37]$$

where the input-output intercorrelation appears:

$$\gamma_{yx}(\tau) = E\left(y(t)x^*(t-\tau)\right) \quad [1.38]$$

1.2.2.3. *Higher order representations*

We are moving away (timidly) from this 2nd order representation for the last few years, because, as has just been explained, it is consistent with a representation of *power* characteristics of the signal. In these representations, the concepts of phase are lost, and the *non-Gaussian* nature of the signals is not clearly demonstrated. These two aspects are among the main motivations to increase the order of representations, leading to *higher order* representations (means greater than 2). We will refer to [LAC 97] and [NIK 93] on these fundamental points. What is important for this book is to note that spectral representations are an essential tool for estimation in this field of *higher order statistics* (HOS).

If we limit this short presentation to null mean, real and stationarity signals of the considered order, the cumulants of the 3rd and 4th order have the following expression, versus the moments:

$$\begin{aligned} C_{xxx}(\tau_1,\tau_2) &= E_{xxx}(\tau_1,\tau_2) = E\big(x(t)x(t-\tau_1)x(t-\tau_2)\big) \\ C_{xxxx}(\tau_1,\tau_2,\tau_3) &= E_{xxxx}(\tau_1,\tau_2,\tau_3) \\ &\quad -\gamma_{xx}(\tau_1)\gamma_{xx}(\tau_2-\tau_3) \\ &\quad -\gamma_{xx}(\tau_2)\gamma_{xx}(\tau_1-\tau_3) \\ &\quad -\gamma_{xx}(\tau_3)\gamma_{xx}(\tau_1-\tau_2) \end{aligned} \qquad [1.39]$$

Using these quantities, we define *multi-spectra* by multidimensional Fourier transform of cumulants:

$$S_{xxx\ldots x}(f_1,\ldots,f_{n-1}) = FT\big(C_{xxx\ldots x}(\tau_1,\ldots,\tau_{n-1})\big) \qquad [1.40]$$

These quantities represent a generalization of PSDs, in the sense that the invariant linear processing (of input $x(t)$ and output $y(t)$) has the advantage of simple formulations. For example, a generalization of Wiener-Lee equations is given by

$$\begin{aligned} S_{yyy\ldots y}(f_1,\ldots,f_{n-1}) = S_{xxx\ldots x}(f_1,\ldots,f_{n-1}).\hat{h}(f_1)\ldots.\hat{h}(f_{n-1}) \\ .\hat{h}(f_1+f_2+\ldots+f_{n-1}) \end{aligned} \qquad [1.41]$$

The multispectral analysis is one of the methods which allows the estimation of HOS. We will, however, note that we do not have results in the form of multispectra as general and elegant as those that are given on p. 14 for the PSDs.

1.3. Spectral analysis: position of the problem

In the previous two sections, we have seen that the characteristics that carry the information in a signal can be extremely diverse. Changing to a transformed domain is a common practice, as in this domain the information "relevant" to the signal (or deemed to be so) is more visible than in the signal itself. In addition, a certain number of processes can be written more clearly in the transformed domain. This is the case with invariant linear processes, whose expression is very compact in the frequency domain.

We are entitled to wonder about the relevance of this omnipresence of the Fourier transform, and its almost systematic use: sclerosis linked to professional habits, or deserved success?

One of the reasons for the success of the kernel $e^{-j2\pi ft}$ (or its equivalent with f complex, the Laplace transform) is that this family of special functions is one of the eigenfunctions of invariant linear systems. When we analyze such systems, it is legitimate to project (break down) the signals on this basis, the linearity of the system ensuring individual processing of each component. Fundamentally, this property is at the root of spectral equations that are as useful as the Wiener-Lee relations.

Outside the invariant linear framework, the Fourier transform with the exponential kernel is not justified as much, and other kernels may be better suited: the linear transforms on Gallois fields in base 2 justify the use of binary kernel transforms, the search for polynomial modulation laws requires higher order Wigner-Ville representations, the highlighting of translations and time dilations induces wavelet representations, etc.

The following chapters deal almost exclusively with the exponential kernel, but the informed reader could easily generalize most of the concepts and results to other kernels, by rather simple transpositions of concepts of frequency resolution, estimation variance, bias-variance trade off, etc.

However, the importance of the exponential kernel for applications is practically a fact of life, the other representations sharing a small portion of applications. This justifies the bias of the current book which is to center the discussion on *spectral analysis with the Fourier kernel*, rather than make a presentation with a more general class of kernels $\psi(t, u)$ and to deduce from this the properties of Fourier analysis: as a result the book loses elegance, but no doubt gains in clarity and pedagogy.

In addition, the majority of chapters is devoted to 2nd order spectral analysis, for the same reasons of extent of the fields of application. However, the extension to multispectra is quite obvious, and the books already listed show that multispectral analysis borrows everything from 2nd order analysis.

In the very large set of spectral analysis methods, we discern *non-parametric* methods, the subject of Chapter 5, which make very few hypotheses on the signal: these hypotheses are essentially the existence of Fourier transforms of the analyzed quantities (the signal itself in the deterministic context, 2nd order moments in random context), and most often the stationarity (strict or local) of moments in random context.

Parametric methods, the subject of Chapters 4 and 6, are based on a more restrictive *a priori* assumption: we suppose that we know a behavioral model of the signal, general model that will be adjusted to the physical signal using a set of parameters. This approach is summarized in Chapter 4. The relevant spectral quantities can be expressed in terms of this set of parameters, which helps the following strategy: we implement parameter estimators, which are used to establish in a second step spectral estimators, thus as by-products of the initial parametric estimation.

A very special case is that of signals made up of a *finite set of sinusoidal functions*, which are deterministic or random, possibly polluted by a random additive phenomenon. This very special framework received great attention from the signal and image processing community, as it helps model a large set of practical situations. We have devoted a large amount of space to it (see sections 4.3 and 6.2.3, and Chapter 8).

The tools necessary for the understanding of developments of Parts 2 and 3 are summarized in the following chapters of this first part.

1.4. Bibliography

[AHM 75] AHMED N., RAO K.R., *Orthogonal Transform for Digital Signal Processing,* Springer Verlag, 1975.

[DUV 91] DUVAUT P., *Traitement du signal*, Hermès, Paris, 1991.

[GAR 89] GARDNER W., *Introduction to Random Processes*, McGraw-Hill, 1989.

[HAR 69] HARMUTH H.F., *Transmission of Information by Orthogonal Functions*, Springer Verlag, 1969.

[HLA 05] HLAWATSCH R., AUGER R., OVARLEZ J.-P., *Temps-fréquence: concepts et outils,* IC2 series, Hermès Science, Paris, 2005.

[LAC 97] LACOUME J.-L., AMBLARD P.-O., COMON P., *Statistiques d'ordre supérieur pour le traitement du signal,* Masson, 1997.

[LAC 00] LACAZE B., *Processus aléatoires pour les communications numériques,* Hermès, Paris, 2000.

[NIK 93] NIKIAS C., PETROPULU A., *Higher-Order Spectra Analysis,* Prentice Hall, 1993.

[OPP 75] OPPENHEIM A., SCHAFFER R., *Digital Signal Processing,* Prentice Hall, 1975.

Chapter 2

Digital Signal Processing

2.1. Introduction

A continuous time deterministic signal $x(t)$, $t \in \Re$ is by definition a function of $\Re$ in C:

$$x : \Re \to C$$
$$t \mapsto x(t)$$

where the variable t is the time. In short, we often talk about a continuous signal, even if the considered signal is not continuous in the usual mathematical sense. For example, we can state displacement, speed and acceleration signals in mechanics, voltage and current signals in electricity, biomedical signals (electrocardiogram, electroencephalogram, electromyogram, etc.), temperature signals, etc.

A discrete time deterministic signal $x(k)$, $k \in Z$ is, by definition, a *series* of complex numbers:

$$x = \left(x(k)\right)_{k \in Z}$$

In short, we often refer to discrete signals. As an example of a discrete time signal, we can state the sunset time according to the day. Generally, the considered signals, whether they are continuous or discrete, have real values, but the generalization given here of complex signals poses no theoretical problem.

Chapter written by Éric LE CARPENTIER.

The spectral analysis of deterministic signals consists of a decomposition based on simpler signals (sine curves for example), similar to the way a point is marked in space using its 3 coordinates. It is thus necessary to define an inner product, which is used to measure the projection of a signal on a basic element.

Let x and y be two signals; their inner product $\langle x, y\rangle$ is defined continuously and discretely respectively by:

$$\langle x,y\rangle = \int_{-\infty}^{+\infty} x(t)\,y^*(t)\,dt \qquad \langle x,y\rangle = \sum_{k=-\infty}^{+\infty} x(k)\,y^*(k) \qquad [2.1]$$

The energy of a signal x is defined by the inner product $\langle x, x\rangle$. The set of signals (with continuous or discrete time) with finite energy, along with the inner product defined above, constitutes a vector space. Very often, the basis elements selected (such as the exponential basis of the Fourier transform) do not verify the finite energy property. A rigorous mathematical processing requires the knowledge of the theory of distributions; we will be content here with an intuitive introduction.

In the case of periodic signals with a known continuous period T or discrete period N, the inner product has the following form:

$$\langle x,y\rangle = \frac{1}{T}\int_{0}^{T} x(t)\,y^*(t)\,dt \qquad \langle x,y\rangle = \frac{1}{N}\sum_{k=0}^{N-1} x(k)\,y^*(k) \qquad [2.2]$$

Thus the inner product $\langle x, x\rangle$ is the power of the periodic signal. The set of periodic signals (with continuous or discrete time) with finite power, along with the inner product defined above, constitute a vector space.

These concepts are dealt with in greater detail in section 2.2.2, particularly in the case of the Fourier transform, in honor of the French mathematician J.B. Fourier (1768-1830), which consists in taking cisoid functions as basis vectors. We will present beforehand some functions and series required for the development of this transform.

2.2. Transform properties

2.2.1. *Some useful functions and series*

The unit constant $1_{\Re}$ is a function that is always equal to 1; for all t:

$$1_{\Re}(t) = 1 \qquad [2.3]$$

The continuous time cisoid of amplitude $a > 0$, frequency $f \in \Re$ and initial phase $\phi \in \Re$ is defined by:

$$t \mapsto a\, e^{j(2\pi f\, t+\phi)} \qquad [2.4]$$

The unit step U is the zero function for the negative instants, and unit for the positive instants. It thus has a discontinuity at 0. The value at 0 is not important. However, we will adopt the following convention: if a signal x has a discontinuity at t_0, then $x(t_0)$ is equal to the arithmetic mean of the left $x(t_0 -) = \lim_{t \uparrow t_0} x(t)$ and right limits $x(t_0 +) = \lim_{t \downarrow t_0} x(t)$. With this convention we thus obtain:

$$u(t)\begin{cases} 0 & \text{if } t < 0 \\ \frac{1}{2} & \text{if } t = 0 \\ 1 & \text{if } t > 0 \end{cases} \qquad [2.5]$$

The rectangular window $1_{t_0 t_1}$ for all t, is defined by:

$$1_{t_0,t_1}(t) = \begin{cases} 1 & \text{if } t_0 < t < t_1 \\ \frac{1}{2} & \text{if } t = t_0 \text{ or } t = t_1 \\ 0 & \text{otherwise} \end{cases} \qquad [2.6]$$

To define the *Dirac delta function* rigorously, knowledge about the mathematical theory of distributions is required. We will be content here with an intuitive introduction. The Dirac delta function δ can be defined as the limit of the function with unit integral $\frac{1}{2T} 1_{-T,T}$ when $T \geq 0$. We thus obtain:

$$\begin{cases} \delta(t) = 0 & \text{if} \quad t \neq 0 \\ \delta(0) = +\infty \end{cases} \quad \text{with the condition} \quad \int_{-\infty}^{+\infty} \delta(t)\, dt = 1 \qquad [2.7]$$

If we integrate the Dirac delta function, we immediately observe that for all t:

$$\int_{-\infty}^{t} \delta(u)\, du = U(t) \qquad [2.8]$$

The Dirac delta function can be considered as the derivative of the unit step:

$$\delta = \frac{dU}{dt} = \dot{U} \qquad [2.9]$$

If we multiply the Dirac delta function δ by a number a, we obtain the Dirac delta function $\alpha\delta$ of weight $\alpha\left(\alpha\delta(t)=0 \text{ if } t\neq 0, \alpha\delta(0)=\infty, \int_{-\infty}^{+\infty}\alpha\delta(t)\,dt=\alpha\right)$.

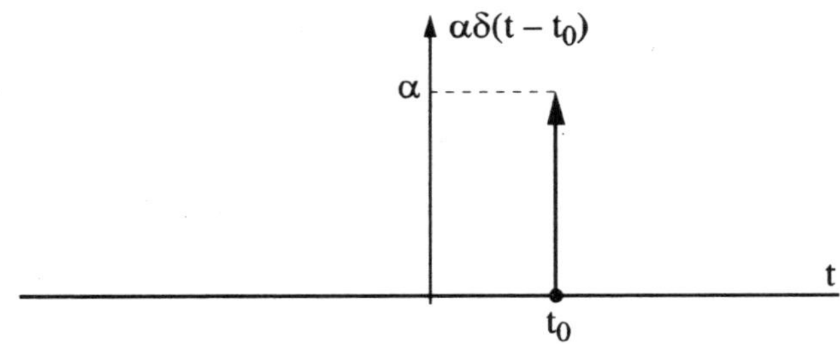

Figure 2.1. *Weighted Dirac delta function at t_0*

The weighted Dirac delta function at t_0, that is $t\mapsto\alpha\delta(t-t_0)$, is generally represented by an arrow centered on t_0 of height proportional to α (Figure 2.1).

The Dirac delta function thus helps generalize the concept of derivative to the signals presenting points of discontinuity, for which the derivative is not defined in the usual sense and at these points is equal to a pulse whose weight is equal to the jump in the discontinuity. The Dirac delta function verifies the following property; for all signals x, and all t_0:

$$\int_{-\infty}^{+\infty} x(t)\,\delta(t-t_0)\,dt = x(t_0) \qquad [2.10]$$

This property is quite natural in the case where the function x is continuous at t_0. We will accept that it remains exact if the convention on the value of a signal at these points of discontinuity is respected. It can be generalized to the ℓth derivative $\delta^{(\ell)}$ of the Dirac delta function; by induction, integrating by parts we obtain for all $\ell\geq 0$ and all t:

$$\int_{-\infty}^{+\infty} x(t)\,\delta^{(\ell)}(t-t_0)\,dt = (-1)^{\ell}\,x^{(\ell)}(t_0) \qquad [2.11]$$

Moreover, time scaling leads to the following property; for all t, and for all non-zero real numbers α:

$$\delta(\alpha t)=\frac{1}{|\alpha|}\delta(t) \qquad [2.12]$$

Finally, we will accept that the Dirac delta function can also be expressed by the following integral formulation:

$$\delta(t) = \int_{-\infty}^{+\infty} e^{j2\pi f t} df \qquad [2.13]$$

The Dirac comb of period T, Ξ_T is a periodized version of Dirac delta functions, defined for all t by:

$$\Xi_T(t) = \sum_{k=-\infty}^{+\infty} \delta(t - kT) \qquad [2.14]$$

We will accept the following formulation in the form of a series; for all t:

$$\Xi_T(t) = \frac{1}{T} \sum_{\ell=-\infty}^{+\infty} e^{j2\pi \frac{\ell}{T} t} \qquad [2.15]$$

The sine cardinal is defined for all t by:

$$\operatorname{sinc}(t) = \begin{cases} \frac{\sin(t)}{t} & \text{if } t \neq 0 \\ 1 & \text{if } t = 0 \end{cases} \qquad [2.16]$$

The value at 0 is obtained by continuity extension. This function (Figure 2.2) becomes zero for all t multiples of -n, except 0.

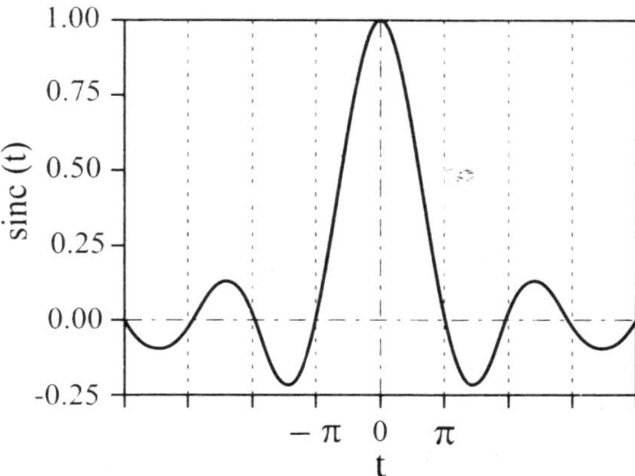

Figure 2.2. *Sine cardinal* sinc

The Dirichlet function (or Dirichlet's kernel, or periodic sinc function) diric_N, parameterized by the integer $N \geq 1$, is defined for all t by:

$$\text{diric}_N(t) = \begin{cases} \dfrac{\sin(\pi\, t\, N)}{N \sin(\pi\, T)} & \text{if } t \notin Z \\ (-1)^{t\,(N-1)} & \text{if } t \in Z \end{cases} \qquad [2.17]$$

The value at the integer abscissa is obtained by continuity extension; it is always equal to 1 if N is odd, (–1)* if N is even. The Dirichlet's function is even and periodic with period 1 if N is odd; it is even, periodic with period 2 and symmetrical in relation to the point $\left(\frac{1}{2}, 0\right)$ if N is even (Figure 2.3). Thus, in all cases, its absolute value is even and periodic with period 1. It is zero for all non-integer t multiple of $\frac{1}{N}$. The arch of this function centered on 0 is called the main lobe, the others are secondary lobes. The main lobe gets narrower and the secondary lobes' amplitude gets smaller as N increases.

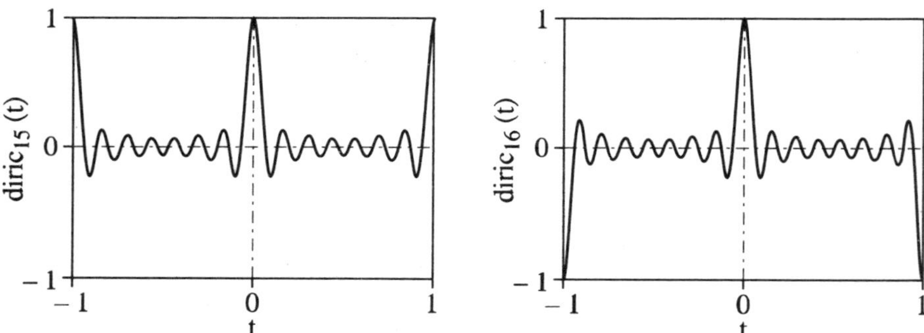

Figure 2.3. *Dirichlet functions* diric_{15} *and* diric_{16}

The unit series 1_Z is the series equal to 1; for all k:

$$1_Z(k) = 1 \qquad [2.18]$$

The discrete time cisoid of amplitude a, frequency ν and initial phase ϕ is defined by:

$$\left(a\, e^{j(2\pi\, \nu\, k + \phi)}\right)_{k \in Z} \qquad [2.19]$$

We immediately notice that the frequency of the cisoid is defined modulo 1.

The rectangular window $1_{k_0,k_1}$ is defined for all *k* by:

$$1_{k_0,k_1}(k)=\begin{cases}1 & \text{if } k_0 \le k \le k_1 \\ 0 & \text{otherwise}\end{cases}$$

The Kronecker δ *sequence* is defined for all *k* by:

$$\begin{cases}\delta(k)=0 & \text{if } k \neq 0 \\ \delta(0)=1 & \end{cases} \quad [2.20]$$

following property; for all signals *x*, and for all k_0:

$$\sum_{k=-\infty}^{+\infty} x(k)\,\delta(k-k_0)=x(k_0) \quad [2.21]$$

Finally, it can be expressed by the following integral formulation:

$$\delta(k)=\int_{-\frac{1}{2}}^{+\frac{1}{2}} e^{j2\pi\,v\,k}\,dv \quad [2.22]$$

The unit comb of period *N*, Ξ_N is a periodized version of the Kronecker sequence, defined for all *k* by:

$$\Xi_N(k)=\sum_{\ell=-\infty}^{+\infty} \delta(k-\ell N) \quad [2.23]$$

The unit comb can also be expressed in the following manner; for all *k*:

$$\Xi_N(k)=\frac{1}{N}\sum_{\ell=0}^{N-1} e^{j2\pi\frac{\ell}{N}k} \quad [2.24]$$

2.2.2. *Fourier transform*

The Fourier transform $\hat{x}(f)$ of a continuous time signal *x*(*t*) is a complex value function of the form $\hat{x}: f \mapsto \hat{x}(f)$ with a real variable, defined for all *f* by:

$$\hat{x}(f)=\int_{-\infty}^{+\infty} x(t)\,e^{-j2\pi f\,t}\,dt \quad [2.25]$$

At the outset let us note that if the variable t is consistent in relation to time, then the variable f is consistent in relation to a frequency. We will accept that the Fourier transform of a signal is defined (i.e. the above integral converges) if the signal is of finite energy.

The Fourier transform does not lead to any loss of information. In fact, knowing $\hat{x}(f)$, we can recreate $x(t)$ from the following inversion formula; for all t:

$$x(t) = \int_{-\infty}^{+\infty} \hat{x}(f)\, e^{j2\pi f\, t} df \qquad [2.26]$$

We demonstrate this theorem by directly calculating the above integral and using the formulae [2.10] and [2.13]:

$$\begin{aligned}\int_{-\infty}^{+\infty} \hat{x}(f)\, e^{j2\pi f\, t} df &= \int_{-\infty}^{+\infty} \int_{-\infty}^{+\infty} x(u)\, e^{j2\pi f\, u} du\, e^{j2\pi f\, t} df \\ &= \int_{-\infty}^{+\infty} x(u) \left[\int_{-\infty}^{+\infty} e^{j2\pi f(t-u)} df \right] du \\ &= \int_{-\infty}^{+\infty} x(u)\, \delta(t-u)\, du \\ &= x(t)\end{aligned}$$

which concludes the demonstration.

For a periodic continuous time signal of period T, the Fourier transform $\hat{x}(f)$ is a series of pulses with $\frac{1}{T}$ spacing whose weight can be calculated using its Fourier series decomposition $\hat{x}(\ell)$:

$$\hat{x}(f) = \sum_{\ell=-\infty}^{+\infty} \hat{x}(\ell)\, \delta\left(f - \frac{\ell}{T}\right) \qquad [2.27]$$

with:

$$\hat{x}(\ell) = \frac{1}{T} \int_{0}^{T} x(t)\, e^{-j2\pi \frac{\ell}{T} t} dt \qquad [2.28]$$

The inverse transform can be written as:

$$x(t) = \sum_{\ell=-\infty}^{+\infty} \hat{x}(\ell) e^{j2\pi \frac{\ell}{T} t} \qquad [2.29]$$

To demonstrate this property, let us calculate the Fourier transform $\hat{x}(f)$; for all f:

$$\hat{x}(f) = \int_{-\infty}^{+\infty} x(t)\, e^{-j2\pi f\, t} dt = \sum_{\ell=-\infty}^{+\infty} \int_{\ell T}^{(\ell+1)T} x(t)\, e^{-j2\pi f\, t} dt$$

Thus, by changing the variable $u = t - \ell T$:

$$\hat{x}(f) = \sum_{\ell=-\infty}^{+\infty} \int_0^T x(u)\, e^{-j2\pi f(u+\ell T)} du$$

$$= \left(\sum_{\ell=-\infty}^{+\infty} e^{-j2\pi f \ell\, T} \right) \left(\int_0^T x(u) e^{-j2\pi f\, u} du \right)$$

We find in the left-hand term, the Dirac comb $\Xi_{1/T}$. Thus:

$$\hat{x}(f) = \frac{1}{T} \Xi_{1/T}(f) \int_0^T x(u)\, e^{-j2\pi f\, u} du$$

$$= \frac{1}{T} \left(\sum_{\ell=-\infty}^{+\infty} \delta\left(f - \frac{\ell}{T} \right) \right) \left(\int_0^T x(u) e^{-j2\pi f\, u} du \right)$$

$$= \sum_{\ell=-\infty}^{+\infty} \left(\frac{1}{T} \int_0^T x(u) \delta\left(f - \frac{\ell}{T} \right) e^{-j2\pi f\, u} du \right)$$

For all f and for all $\ell, \delta\left(f - \frac{\ell}{T}\right) e^{-j2\pi fu} = \delta\left(f - \frac{\ell}{T}\right) e^{-j2\pi \frac{\ell}{T} u}$. Therefore:

$$\hat{x}(f) = \sum_{\ell=-\infty}^{+\infty} \left(\frac{1}{T} \int_0^T x(u) e^{-j2\pi \frac{\ell}{T} u} du \right) \delta\left(f - \frac{\ell}{T} \right) = \sum_{\ell=-\infty}^{+\infty} \hat{x}(\ell)\, \delta\left(f - \frac{\ell}{T} \right)$$

By the inverse Fourier transform, we obtain the formula [2.29], which concludes the demonstration.

The Fourier transform $\hat{x}(\nu)$ of a discrete time signal $x(k)$ is a function of the form:

$$\begin{aligned} \hat{x} : \Re &\rightarrow C \\ \nu &\mapsto \hat{x}(\nu) \end{aligned}$$

defined for all ν by:

$$\hat{x}(\nu) = \sum_{k=-\infty}^{+\infty} x(k) e^{-j2\pi\,\nu\,k} \qquad [2.30]$$

We will accept that the Fourier transform of a discrete time signal is defined (i.e. the series mentioned above converges) if the signal is of finite energy. It is periodic with period 1.

The Fourier transform of discrete time signals can be linked to the Fourier transform of continuous time signals in the following way. Let x_I be the continuous time signal obtained from x in the following manner, where T is a positive real number:

$$x_{\mathrm{I}}(t) = T \sum_{k=-\infty}^{+\infty} x(k)\delta(t-kT)$$

x_I is thus zero everywhere except at instants $k\,T$ multiples of the period T where it has a pulse of weight $T\,x(k)$. Using the formula [2.10] we easily obtain that the Fourier transform $\hat{x}(\nu)$ of a discrete time signal $x(k)$ is none other than the Fourier transform $\hat{x}_{\mathrm{I}}(f)$ of a continuous time signal $x_I(t)$, with a change in scale and rounded to a factor, the result being independent of T; for all ν:

$$\hat{x}(\nu) = \frac{1}{T}\hat{x}_{\mathrm{I}}\left(\frac{\nu}{T}\right) \qquad [2.31]$$

The Fourier transform does not result in any loss of information. In fact, knowing $\hat{x}(\nu)$, we can recreate $x(k)$ from the following inversion formula (obtained from a demonstration similar to the continuous case); for all values k:

$$x(k) = \int_{-\frac{1}{2}}^{+\frac{1}{2}} \hat{x}(\nu) e^{j2\pi\,\nu\,k} d\nu \qquad [2.32]$$

For a periodic discrete time signal of period N, the Fourier transform $\hat{x}(\nu)$ is a series of comb pulses with $\frac{1}{N}$ spacing whose weight can be calculated using its decomposition into the discrete Fourier transform $\hat{x}(\ell)$:

$$\hat{x}(\nu) = \sum_{\ell=0}^{N-1} \hat{x}(\ell)\Xi_1\left(\nu - \frac{\ell}{N}\right) \qquad [2.33]$$

with:

$$\hat{x}(\ell) = \frac{1}{N}\sum_{k=0}^{N-1} x(k)e^{-j2\pi\frac{\ell k}{N}} \qquad [2.34]$$

The inverse transform can be written as:

$$x(k) = \sum_{\ell=0}^{N-1} \hat{x}(\ell)e^{j2\pi\frac{\ell n}{N}} \qquad [2.35]$$

2.2.3. *Fundamental properties*

Examination of the formulae [2.26], [2.29], [2.32] and [2.35] shows that the different forms of the Fourier transform help break down the signal in the form of a sum of cisoids. A discrete time signal has a periodic transform. Inversely, a periodic signal is broken down into a discrete (countable) set of cisoids. The modulus of the transform provides information on the amplitude of these cisoids and is called the amplitude spectrum. The phase provides information on the initial phase of these cisoids and is called the phase spectrum.

The Fourier transform verifies the linearity property: let x and y be two signals and let a be a complex number. Then:

$$\begin{cases} \widehat{x+y} = \hat{x} + \hat{y} \\ \widehat{ax} \quad = a\hat{x} \end{cases} \qquad [2.36]$$

The Fourier transform of the conjugate signal x^* is the Fourier transform of the returned signal x and conjugated. The Fourier transform of the returned signal x^- is the Fourier transform of the returned signal x.

$$\widehat{x^*} = (\hat{x})^{-*} \qquad [2.37]$$

$$\widehat{x^{-}} = \hat{x}^{-} \qquad [2.38]$$

We can deduce the properties of symmetry of the transform of particular signals represented in Table 2.1.

Signal		Transform
Real number	$x = x^*$	Even real part Odd imaginary part
Imaginary	$x = x^*$	Odd real part Even imaginary part
Even	$x = x^{-}$	Even
Odd	$x = x^{-}$	Odd
Even real parts Odd imaginary part	$x = x^*$	Real
Odd real parts Even imaginary part	$x = -x^*$	Imaginary

Table 2.1. *Symmetry of the Fourier transform*

2.2.4. *Convolution sum*

A particular operation between functions or series is the convolution sum ⊗. Let $x(t)$ and $y(t)$ be two functions with a real variable; their convolution sum $x \otimes y(t)$ is defined for all t by:

$$x \otimes y(t) = \int_{-\infty}^{+\infty} x(\tau) y(t-\tau) d\tau \qquad [2.39]$$

Similarly, let $x(k)$ and $y(k)$ be two series; their convolution sum $x \otimes y(k)$ is defined for all k by:

$$x \otimes y(k) = \sum_{m=-\infty}^{+\infty} x(m) y(k-m) \qquad [2.40]$$

When two functions $x(t)$ and $y(t)$ are periodic, of similar period T (or if they are both of finite support [0, T]), we can also define their circular convolution sum $x \otimes y(t)$, which is a periodic function of period T such that, for all values of t:

$$x \otimes y(t) = \frac{1}{T} \int_{0}^{T} x(\tau) y((t-\tau) \bmod T) d\tau \qquad [2.41]$$

Similarly, for two periodic series $x(k)$ and $y(h)$ with the same period N (or if they are both of finite support {0,..., $N-1$}), their circular convolution sum $x \otimes y(k)$ is periodic of period N, and defined for all values of n by:

$$x \otimes y(k) = \frac{1}{N} \sum_{m=0}^{N-1} x(m)\, y((k-m) \bmod N) \qquad [2.42]$$

The convolution sum and the circular convolution sum verify the commutative and associative properties and the identity element is:

$$\begin{cases} \delta(t) & \text{for the functions} \\ \delta(k) & \text{for the series} \\ T\Xi_T(t) & \text{for the functions of period } T \\ N\Xi_N(k) & \text{for the series of period } N \end{cases} \qquad [2.43]$$

In addition, by noting $x_{t0}(t)$ the delayed function $x(t)$ by $t_0 \left(x_{t0}(t) \triangleq x(t - t_0) \right)$ we can easily show:

$$\begin{cases} x_{t_0} \otimes y(t) = x \otimes y_{t_0}(t) & \text{for the functions} \\ x_{k_0} \otimes y(k) = x \otimes y_{k_0}(k) & \text{for the series} \\ x_{t_0} \otimes y(\ell) = x \otimes y_{t_0}(\ell) & \text{for the functions of period } T \\ x_{k_0} \otimes y(\ell) = x \otimes y_{k_0}(\ell) & \text{for the series of period } N \end{cases} \qquad [2.44]$$

In particular, the convolution of a function or a series with the delayed identity element delays it by the same amount.

We can easily verify that the Fourier transform of a convolution sum or circular convolution sum is the product of transforms:

$$\begin{cases} \widehat{x \otimes y}(f) = \hat{x}(f)\,\hat{y}(f) & \text{Fourier transform (continuous time)} \\ \widehat{x \otimes y}(\nu) = \hat{x}(\nu)\,\hat{y}(\nu) & \text{Fourier transform (discrete time)} \\ \widehat{x \otimes y}(\ell) = \hat{x}(\ell)\,\hat{y}(\ell) & \text{Fourier series (period } T) \\ \widehat{x \otimes y}(\ell) = \hat{x}(\ell)\,\hat{y}(\ell) & \text{Discrete Fourier transform (period } N) \end{cases} \qquad [2.45]$$

Inversely, the transform of a product is a convolution sum:

$$\begin{cases} \widehat{xy}(f) = \hat{x} \otimes \hat{y}(f) & \text{Fourier transform (continuous time)} \\ \widehat{xy}(\nu) = \hat{x} \otimes \hat{y}(\nu) & \text{Fourier transform (discrete time)} \\ \widehat{xy}(\ell) = \hat{x} \otimes \hat{y}(\ell) & \text{Fourier series (period } T\text{)} \\ \widehat{xy}(\ell) = N\, \hat{x} \otimes \hat{y}(\ell) & \text{Discrete Fourier transform (period } N\text{)} \end{cases} \quad [2.46]$$

2.2.5. *Energy conservation (Parseval's theorem)*

The energy of a continuous time signal $x(t)$ or a discrete time signal $x(k)$ can be calculated by integrating the square of the Fourier transform modulus $\hat{x}(\nu)$ or its transform in standardized frequency $\hat{x}(\nu)$:

$$\langle x, x \rangle = \langle \hat{x}, \hat{x} \rangle \quad [2.47]$$

The function or the series $|\hat{x}|^2$ is known as the energy spectrum, or energy spectral density of a signal x, because its integral (or its summation) gives the energy of the signal x.

The power of a periodic signal of continuous period T, or discrete period N, can be calculated by the summation of the square of the decomposition modulus in the Fourier series or discrete Fourier transform. For the Fourier series, this is written as:

$$\langle x, x \rangle = \langle \hat{x}, \hat{x} \rangle \quad [2.48]$$

and for the discrete Fourier transform as:

$$\langle x, x \rangle = N \langle \hat{x}, \hat{x} \rangle \quad [2.49]$$

To sum up, by recalling the definition of the inner product, Parseval's theorem can be written as:

– continuous time signals: $\int_{-\infty}^{+\infty} |x(t)|^2 dt = \int_{-\infty}^{+\infty} |\hat{x}(f)|^2 df$

– T-periodic signals: $\frac{1}{T} \sum_{0}^{T} |x(t)|^2 dt = \sum_{\ell=-\infty}^{+\infty} |\hat{x}(\ell)|^2$

– discrete time signals: $\sum_{k=-\infty}^{+\infty} |x(k)|^2 = \int_{-\frac{1}{2}}^{+\frac{1}{2}} |\hat{x}(v)|^2 \, dv$

– N-periodic signals: $\frac{1}{N}\sum_{k=0}^{N-1} |x(k)|^2 = \sum_{\ell=0}^{N-1} |\hat{x}(\ell)|^2$

Let us demonstrate this property in the case of continuous time signals. Let x^- be the returned signal x (for all values of t, $x^-(t) = x(-t)$). By combining the formulae [2.37] and [2.38], we obtain:

$$\widehat{x^{-*}}(f) = [\hat{x}(f)]^*$$

Thus:

$$\widehat{x \otimes x^{-*}}(f) = \hat{x}(f)\widehat{x^{-*}}(f) = \hat{x}(f)[\hat{x}(f)]^* = [\hat{x}(f)]^2$$

thus, by inverse Fourier transform calculated at $t = 0$:

$$x \otimes x^{--*}(0) = \int_{-\infty}^{+\infty} |\hat{x}(f)|^2 \, df$$

Moreover:

$$x \otimes x^{-*}(0) = \int_{-\infty}^{+\infty} x(t)x^{-*}(0-t)\,dt = \int_{-\infty}^{+\infty} x(t)x^*(t)\,dt = \int_{-\infty}^{+\infty} |x(t)|^2 \, dt$$

which concludes the demonstration.

2.2.6. *Other properties*

MODULATION – The product of a continuous time signal $x(t)$ by a complex cisoid of frequency f_0 offsets the Fourier transform by f_0:

$$y(t) = x(t)e^{j2\pi f_0 t} \qquad \hat{y}(f) = \hat{x}(f - f_0) \qquad [2.50]$$

The product of a discrete time signal $x(t)$ by a complex cisoid of frequency v_0 offsets the Fourier transform by v_0:

$$y(k) = x(k)e^{j2\pi v_0 k} \qquad \hat{y}(v) = \hat{x}(v - v_0) \qquad [2.51]$$

This property is used in telecommunication systems by modulating the amplitude, f_0 being the frequency of the carrier sine wave.

TIME SHIFTING – *Shifting a continuous time signal* $x(t)$ *by a time* t_0 *(if* $t_0 > 0$*, it consists of a lag; if* $t_0 < 0$*, it consists of a lead) means multiplying the Fourier transform by the cisoid of "frequency"* t_0*:*

$$y(t) = x(t - t_0) \qquad \hat{y}(f) = e^{j2\pi f t_0} \hat{x}(f) \tag{2.52}$$

Shifting the discrete time signal $x(k)$ *by a time* $k_o \in Z$ *means multiplying the Fourier transform b the cisoid of "frequency"* k_0*:*

$$y(k) = x(k - k_0) \qquad \hat{y}(v) = e^{j2\pi v k_0} \hat{x}(v) \tag{2.53}$$

DERIVATIVE – *Let* $x(t)$ *be a continuous time signal of derivative* $\dot{x}(t) = \frac{dx}{dt}(t)$. *Then, for all values of* f*:*

$$\hat{\dot{x}}(f) = j2\pi f \hat{x}(f) \tag{2.54}$$

For a T-periodic signal, whose derivative is also T-periodic, the decomposition into Fourier series of the derivative signal can be written as:

$$\hat{\dot{x}}(\ell) = j2\pi \frac{\ell}{T} \hat{x}(\ell) \tag{2.55}$$

TIME SCALING – *The time dependent expansion by a factor* $a \in \Re$ *of a continuous time signal leads to a frequency contraction by a factor* $\frac{1}{a}$*:*

$$y(t) = x(at) \qquad \hat{y}(f) = \frac{1}{|a|} \hat{x}\left(\frac{f}{a}\right) \tag{2.56}$$

For a discrete time signal, the equivalent property must be used with greater care. Let $a \in Z$*, and let* x *be a zero discrete time signal for all the non-multiple instants of* a*. Consider the signal* y *obtained from the signal* x *by the time scaling by the factor* a*; then, the Fourier transform of* y *is obtained by the contraction of* x*:*

$$y(k) = x(ak) \qquad \hat{y}(v) = \hat{x}\left(\frac{v}{a}\right) \tag{2.57}$$

2.2.7. *Examples*

Dirac delta function

From the formula [2.10], we immediately obtain:

$$\hat{\delta}(f) = 1_{\Re}(f) \qquad [2.58]$$

Continuous time rectangular window

The Fourier transform of the continuous time rectangular window $1_{-T/2,T/2}$ is the sine cardinal function:

$$\hat{1}_{-T/2,T/2}(f) = T\,\mathrm{sinc}(\pi T f) \qquad [2.59]$$

Sine cardinal

The Fourier transform of the sinus cardinal is the rectangular window:

$$\widehat{\mathrm{sinc}}(f) = \pi 1_{-\frac{1}{2\pi},\frac{1}{2\pi}}(f) \qquad [2.60]$$

Unit constant

It is not of finite energy, but nevertheless has a Fourier transform, in terms of the theory of distributions, which is the Dirac delta function:

$$\hat{1}_{\Re}(f) = \delta(f) \qquad [2.61]$$

This result can be easily obtained by exchanging the rôle of the variables t and f in formula [2.13].

Continuous time cisoid

We have the following transformation:

$$x(t) = e^{j2\pi f_0 t} \qquad \hat{x}(f) = \delta(f - f_0)$$

This means that the Fourier transform of the cisoid of frequency f_0 is a pulse centered on f_0. By using the linearity of the Fourier transform, we easily obtain the Fourier transform of a real sine wave, whatever its initial phase; particularly:

$$x(t) = \cos(2\pi f_0 t) \qquad \hat{x}(f) = \frac{1}{2}\left[\delta(f - f_0) + \delta(f - f_0)\right] \qquad [2.62]$$

$$x(t) = \sin(2\pi f_0 t) \qquad \hat{x}(f) = \frac{-j}{2}\left[\delta(f - f_0) - \delta(f - f_0)\right] \qquad [2.63]$$

Dirac comb

The Fourier transform of the Dirac comb is a Dirac comb:

$$\hat{\Xi}_T(f) = \frac{1}{T}\Xi_{1/T}(f) \qquad [2.64]$$

This result is immediately obtained from formula [2.15] and the linearity of the Fourier transform.

Kronecker sequence

From formula [2.21], we immediately obtain:

$$\hat{\delta}(\nu) = 1_{\Re}(\nu) \qquad [2.65]$$

Discrete time rectangular window

The Fourier transform of the gateway $1_{0,N-1}$ is expressed according to Dirichlet's kernel (formula [2.17]):

$$\hat{1}_{0,N-1}(\nu) = N\, e^{-j\pi(N-1)\nu}\, \mathrm{diric}_N(\nu) \qquad [2.66]$$

It is equal to N for all integer values of the abscissa.

Unit series

The Fourier transform of the constant series $1_z(k)$ is the Dirac comb Ξ_1:

$$\hat{1}_Z(\nu) = \Xi_1(\nu) \qquad [2.67]$$

This result is obtained using formula [2.15].

Discrete time cisoid

We have the following transformation:

$$x(k) = e^{j2\pi\,\nu_0\,k} \qquad \hat{x}(f) = \Xi_1(\nu - \nu_0) \qquad [2.68]$$

This means that the Fourier transform of the cisoid of frequency ν_0 is a comb centered on ν_0. Using the linearity of the Fourier transform, we obtain the Fourier transform of real sine waves.

Unit comb

The Fourier transform of the unit comb is a Dirac comb:

$$\hat{\Xi}_N(\nu) = \frac{1}{N}\Xi_{\frac{1}{N}}(\nu) \qquad [2.69]$$

This result is obtained using formula [2.24].

2.2.8. *Sampling*

In the majority of applications, the signal studied is a continuous time signal. The spectral analysis, using a digital calculator, requires the sampling of this signal to obtain a discrete time signal. This sampling must be done in a suitable way, so as not to lose too much information. We will develop later the Shannon condition, a mathematical condition which makes it possible to understand the concept of suitable sampling and its practical implications.

Consider a continuous time signal $x(t)$, sampled at the period T_c, and considered continuous at sampling instants; we thus obtain the discrete time signal $x_e(k)$; for all values of k:

$$x_e(k) = x(kT_c)$$

Knowing the sampled signal $x_e(k)$, is it possible to recreate the original signal $x(t)$? In the spectral domain, it is equivalent to asking the following question: knowing $\hat{x}_e(v)$, can we recreate $\hat{x}(f)$? Let us recall the expression of these two Fourier transforms; for all values of f:

$$\hat{x}(f) = \int_{-\infty}^{+\infty} x(t) e^{-j2\pi f t} dt$$

And for all values of v:

$$\hat{x}_e(v) = \sum_{k=-\infty}^{+\infty} x_e(k) e^{-j2\pi v k}$$

We note that if we bring the integral close to the definition of $\hat{x}$ by the rectangle approximation, we obtain for all values of f:

$$\begin{aligned}\hat{x}(f) &\approx T_c \sum_{k=-\infty}^{+\infty} x(kT_c) e^{-j2\pi f n T_c} \\ &\approx T_c \hat{x}_e(fT_c)\end{aligned}$$

We see that sampling leads to a loss of information. However, under certain conditions, we can exactly recreate the continuous signal from the sampled signal and show that the relation mentioned above is not an approximate one. Consider $x_l(t)$ as a continuous time signal obtained from the sampled signal $x_e(k)$ by:

$$x_1(t) = T \sum_{k=-\infty}^{+\infty} x_e(k) \delta(t - kT_c)$$

By keeping in mind the definition of the Dirac comb Ξ, we obtain:

$$x_1(t) = T_c\, x(t)\, \Xi_{T_c}(t) \qquad [2.70]$$

Thus, by the Fourier transform, using the formulae [2.31], [2.70], [2.46] and [2.64] successively, we have for all values of f:

$$\begin{aligned}\hat{x}_e(fT_c) &= \frac{1}{T_c}\hat{x}_1(f)\\ &= \widehat{x\Xi_{T_c}}(f)\\ &= \hat{x}\otimes\hat{\Xi}_{T_c}(f)\\ &= \frac{1}{T_c}\left(\hat{x}\otimes\Xi_{\frac{1}{T_c}}\right)(f)\end{aligned}$$

Then, using the definition of the convolution sum [2.39], the definition of the Dirac comb [2.14] and the formula [2.10], we obtain:

$$\begin{aligned}\hat{x}_e(fT_c) &= \frac{1}{T_c}\int_{-\infty}^{+\infty}\hat{x}(g)\,\Xi_{\frac{1}{T_c}}(f-g)\,dg\\ &= \frac{1}{T_c}\int_{-\infty}^{+\infty}\hat{x}(g)\sum_{k=-\infty}^{+\infty}\delta\left(f-g-\frac{k}{T_c}\right)dg\\ &= \frac{1}{T_c}\sum_{k=-\infty}^{+\infty}\int_{-\infty}^{+\infty}\hat{x}(g)\,\delta\left(f-g-\frac{k}{T_c}\right)dg\\ &= \frac{1}{T_c}\sum_{k=-\infty}^{+\infty}\hat{x}\left(f-\frac{k}{T_c}\right)\end{aligned}$$

This means that the Fourier transform of the sampled signal approximated to the factor T_c, is the sum of the Fourier transform of the continuous time signal and its translated versions by a multiple of $\frac{1}{T_c}$ We obtain a periodic function of period $\frac{k}{T_c}$; we call this phenomenon the periodization of the spectrum.

For example, let us observe what happens when the Fourier transform of a continuous signal $\hat{x}(f)$ is of finite support $\{f \| f| < f_{\max}\}$ and has the shape of

Figure 2.4(a). If $f_{\max} < \frac{1}{2T_c}$, the shifted copies of the spectrum lead to Figure 2.4(b). The spectrum of the sampled signal is thus represented in Figure 2.4(c). We see that if $f_{\max} > \frac{1}{2T_c}$, the spectrum is not distorted. However, if $f_{\max} > \frac{1}{2T_c}$, the shifted copies of the spectrum lead to Figure 2.4(d). The spectrum of the sampled signal is thus represented in Figure 2.4(e). Thus, if $f_{\max} > \frac{1}{2T_c}$, we notice a distortion of the spectrum close to the Nyquist frequency $\frac{1}{2T_c}$. This phenomenon is called spectrum aliasing. We can thus conclude that, if a continuous signal $x(t)$ of period T_c is sampled such that the Fourier transform $x(f)$ is zero outside the interval $\left[-\frac{1}{2T_c}, \frac{1}{2T_c}\right]$ (this condition is called the Shannon condition), then for all $f \in \left[-\frac{1}{2T_c}, \frac{1}{2T_c}\right]$ we have:

$$\hat{x}(f) = T_c \hat{x}_e(f T_c) \tag{2.71}$$

By the inverse Fourier transform, we can recreate the continuous time signal x by the following integral formulation; for all values of t:

$$\begin{aligned} x(t) &= \int_{-\frac{1}{2T_c}}^{+\frac{1}{2T_c}} \hat{x}(f) e^{j2\pi f t} df \\ &= T_c \int_{-\frac{1}{2T_c}}^{+\frac{1}{2T_c}} \hat{x}_e(f T_c) e^{j2\pi f t} df \\ &= \int_{-\frac{1}{2}}^{+\frac{1}{2}} \hat{x}_e(\nu) e^{j2\pi \frac{\nu}{T_c} t} d\nu \end{aligned} \tag{2.72}$$

Moreover, the formula [2.71] can be written more concisely; if the Shannon condition is verified, then for all values of f:

$$\hat{x}(f) = \hat{x}_e(f T_c) T_c 1_{-\frac{1}{2T_c}, \frac{1}{2T_c}}(f) \tag{2.73}$$

By using the Fourier transform definition, the delay theorem and the formula [2.60], we obtain the reconstruction of the continuous time signal $x(t)$ by the formulation in the form of the following series; for all values of t:

$$x(t) = \sum_{k \in Z} x_e(k) \operatorname{sinc}\left[\pi\left(\frac{t}{T_c} - k\right)\right] \tag{2.74}$$

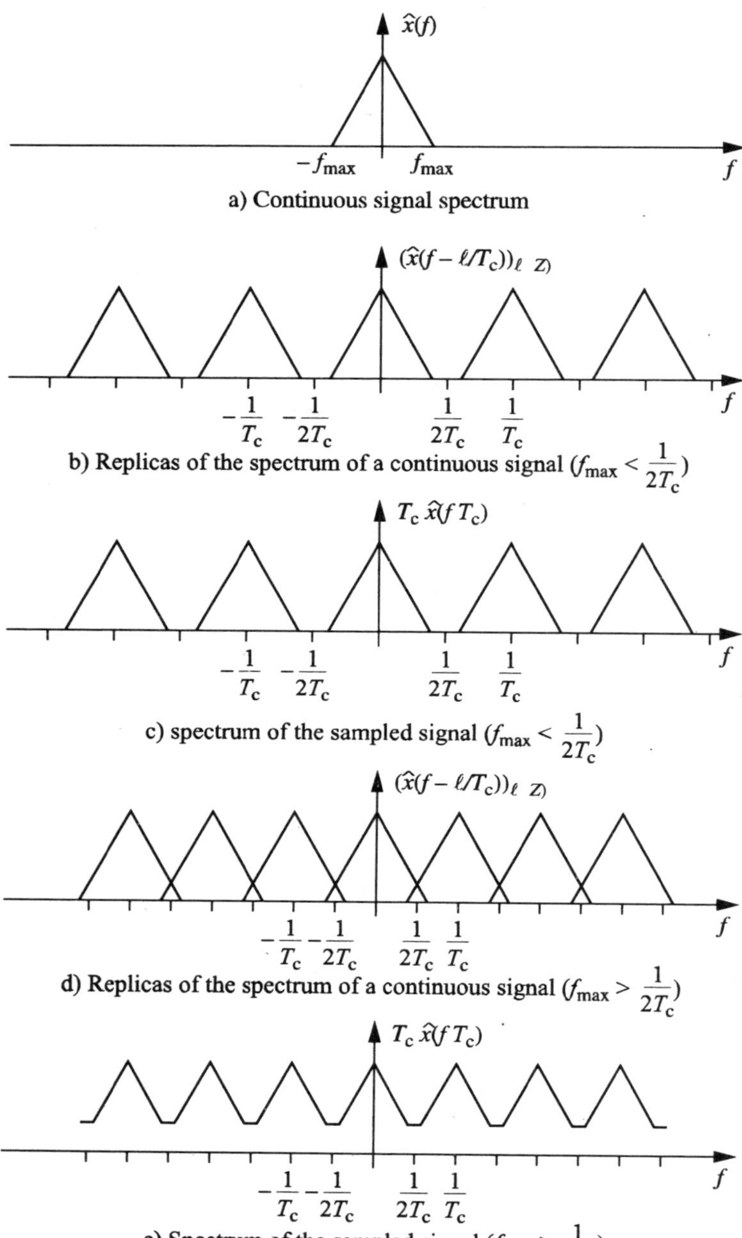

Figure 2.4. *Periodization and spectral aliasing*

To visualize these concepts, let us consider a real sine wave of 70 Hz, sampled at 65 Hz (Figure 2.5(a)). The sampled signal is thus a sine wave of 5 Hz frequency. In fact, the spectrum of amplitude of the continuous time sine wave is even, and consists of two pulses at 70 Hz and -70 Hz (Figure 2.5(b)). To sample correctly, at least two sampling points per period of the continuous time sine wave are necessary. After sampling, we obtain the spectrum as in Figure 2.5(c), which has a peak at 5 Hz.

The spectral aliasing phenomenon also explains why the wheel of a vehicle being filmed seems to rotate slowly in the other direction or even seems stationary when the images are sampled at 24 Hz (25 Hz on television). The spectral aliasing can also be found for example in the tuning of the ignition point of a spark ignition engine motor through a stroboscope.

Generally, one must be careful when a continuous time signal is sampled, so that the Shannon condition is fulfilled, at the risk of seeing the low frequencies generated by the aliasing of high frequencies of the continuous time signal in the sampled signal. In practice, before any sampling, the signal observed must be filtered by a low-pass analog filter, known as *anti-aliasing,* whose cut-off frequency is 2.5 to 5 time less than the sampling frequency.

2.2.9. *Practical calculation, FFT*

In practice, we record a signal $x = \left(x(k)\right)_{k \in Z}$, which has a N finite number of points, that is $\left(x(k)\right)_{0 \le k \le N-1}$. It is thus tempting to approach the Fourier transform $\hat{x}$, that is:

$$\hat{x}(\nu) = \sum_{k=-\infty}^{+\infty} x(k) e^{-j2\pi\nu k}$$

by the finite sum:

$$X(\nu) = \sum_{k=0}^{N-1} x(k) e^{-j2\pi \nu k}$$

This amounts to approaching x by the Fourier transform of the signal x multiplied by the rectangular window $1_{0,N-1}$ i.e. $\widehat{x 1_{0,N-1}}$. We will see the effects and inconveniences of this truncation in section 2.3.

To come back to the calculation of a finite series, it is necessary to sample the frequency axis [0,1[, for example, by using regular sampling of M points in this interval:

$$\nu \in \left\{ \frac{m}{M}, 0 \le m \le M-1 \right\} \qquad [2.75]$$

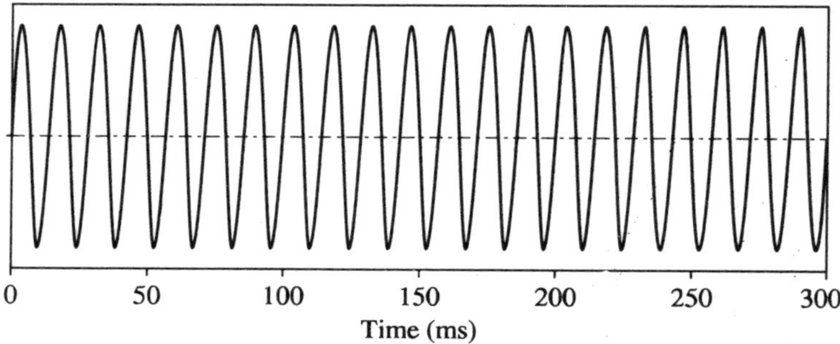

a) Sine wave of 70 Hz (—), sampled at 65 Hz(.)

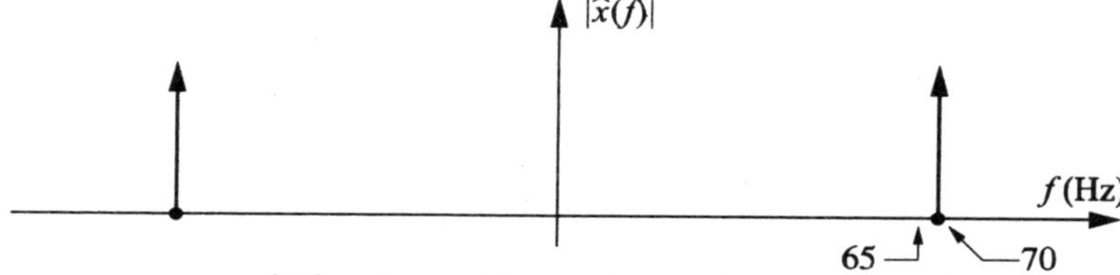

b) Spectrum of the continuous sine wave

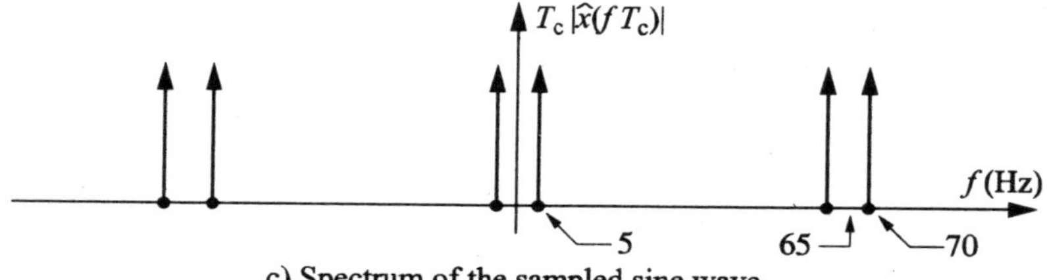

c) Spectrum of the sampled sine wave

Figure 2.5. *Sine wave and spectral folding*

Thus we have, for all values of $m \in \{0,..., M-1\}$, by noting $X_e(m) = X(\frac{m}{M})$:

$$X_e(m) = \sum_{k=0}^{N-1} x(k) e^{-j2\pi \frac{mk}{M}} \qquad [2.76]$$

If $M = N$, we recognize, to the multiplicative factor $\frac{1}{N}$, the operator of the discrete Fourier transform. If $M > N$, we recognize the operator of the discrete transform applied to the series $\left(x(k)\right)_{0 \le k \le N-1}$ completed by $M - N$ zeros (this addition of zeros is known as *zero padding*).

Implicitly, this approximate calculation amounts to periodizing the signal: we calculate the discrete Fourier transform of the N-periodic signal whose one period is the interval of N points available, possibly extended by $M - N$ zeros.

This calculation, which is performed using $M\ (M - 1)$ additions and M^2 multiplications, that is $O(M^2)$ operations, can become very cumbersome for a large M. The FFT algorithms help reduce this calculation complexity, if M is not a prime number. The most widely used algorithm known as Cooley-Tukey in base 2 can reduce this number of calculations to $O\ (M \log_2 M)$ operations and this is only in the case where M is a power of 2. It is based on the following time dependent decimation principle. Let X_{even} (respectively X_{odd}) be the transform of the series of $\frac{M}{2}$ even index points (respectively with odd index) extracted from the series x (to which we have added $M - N$ zeros):

$$X_{\text{even}}(m) = \sum_{k=0}^{\frac{M}{2}-1} x(2k)\, e^{-j2\pi \frac{m\,k}{M/2}}$$

$$X_{\text{odd}}(m) = \sum_{k=0}^{\frac{M}{2}-1} x(2k+1)\, e^{-j2\pi \frac{m\,k}{M/2}}$$

We thus easily obtain for all values of $m \in \{0, \ldots, M - 1\}$, by drawing options from the periodicity of period $\frac{M}{2}$ of X_{even} and X_{odd}:

$$X_{\text{e}}(m) = X_{\text{even}}(m) + e^{-j2\pi \frac{m}{M}} X_{\text{odd}}(m) \qquad [2.77]$$

This calculation consists of an addition and a multiplication, for fixed m. The calculation of the complete series $(X_e(m))_{0 \le m \le M-1}$ thus consists of M additions and M multiplications. The discrete Fourier transform of order M is thus calculated in function of 2 discrete Fourier transforms of order $\frac{M}{2}$ using M additions and M multiplications. By iterating the process, we notice that these 2 discrete Fourier transforms of order $\frac{M}{2}$ are calculated in function of 4 discrete Fourier transforms of order $\frac{M}{4}$ using $2 \times \frac{M}{2}$ additions and $2 \times \frac{M}{2}$ multiplications, that is M additions and M multiplications. This process can be iterated $\log_2 M$ times, until the calculation of M discrete Fourier transforms of the 1st order, which requires no operations. Finally, this calculation thus requires $M \log_2 M$ additions and $M \log_2 M$ multiplications.

If the fact that M is a power of 2 seems very restrictive, it is possible to develop the FFT algorithms in the case where M can be factorized in the form of a product of integer numbers $M = M_1 M_2, \ldots M_p$. Thus we are talking of algorithms in a multiple base, which require $O\left(M\left(M_1 + \cdots + M_p\right)\right)$ operations [BRI 74].

These fast algorithms require the calculation of the discrete Fourier transform in its entirety. It is often preferable to return to a direct calculation if we do not need to calculate this transform only in low frequencies known *a priori*.

2.3. Windows

Consider a continuous time signal $x(t)$ with an unlimited support which we recorded during the time interval $[0, T]$. Thus it is natural to approximate its Fourier transform $\hat{x}(f)$, that is:

$$\hat{x}(f) = \int_{-\infty}^{+\infty} x(t) e^{-j2\pi f t} dt$$

by the finite integral:

$$\int_0^T x(t) e^{-j2\pi f t} dt$$

This amounts to approximating $\hat{x}(f)$ by the Fourier transform of the signal $x(t)$ multiplied by the rectangular window $1_{0,T}(t)$. Let us observe $\widehat{x1_{0,T}}(f)$ the result of this approximation in the case where the signal $x(t)$ is a cisoid of frequency f_0, amplitude a and initial phase ϕ:

$$x(t) = a e^{j(2\pi f_0 t + \phi)} \qquad \hat{x}(v) = a e^{j\phi} \delta(f - f_0)$$

The product is transformed by the Fourier transform into a convolution sum. This convolution along with the Dirac delta function shifted to F_0 thus shifts the Fourier transform of the window to f_0. We thus obtain:

$$\widehat{x1_{0,T}}(f) = a\,e^{j\phi}\,\hat{1}_{0,T}(f - f_0)$$

Thus, in the place of the expected pulse at f_0, we observe the sine cardinal centered on v_0 (see Figure 2.6).

This spectral "leakage" exists even if the signal processed is not a sine wave. In particular, if the signal consists of two sine waves of very different amplitudes, the secondary lobes coming from the sine wave of large amplitude can mask the main lobe coming from the other sine wave. A solution consists in balancing the signal observed using a smoother truncation window, whose Fourier transform is such that the amplitude of the secondary lobes is much smaller compared to the amplitude of the main lobe, than in the case of the rectangular window. We will thus approximate the Fourier transform $x(f)$ of the signal $x(t)$ by $\widehat{xw_{0,T}}(f)$ where $w_{0,T}(t)$ is the zero signal outside the interval [0, T].

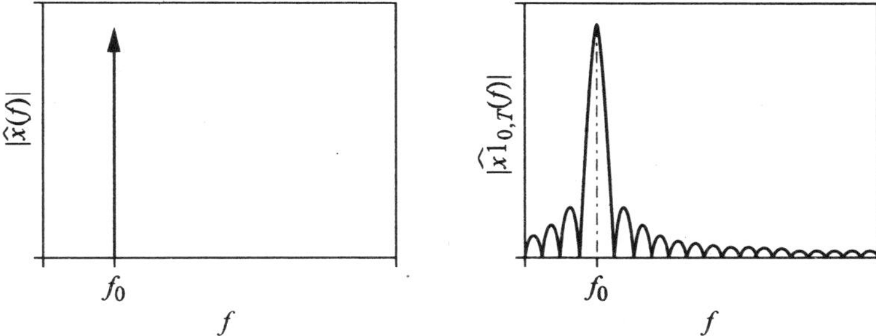

Figure 2.6. *Fourier transform of a sine wave before and after truncation*

Many truncation windows exist in the literature, all symmetrical in relation to the $t = \frac{T}{2}$ axis. Table 2.2 represents the common windows and Table 2.3 their Fourier transform. Usually this transform is represented in decibels, by plotting the frequency response in decibels that is 10 times the logarithm in base 10 of the spectral energy density, or 20 times the logarithm in base 10 of the spectral amplitude density:

$$W_{\text{dB}}(f) = 10\log_{10}\left|\hat{w}_{0,T}(f)\right|^2 = 20\log_{10}\left|\hat{w}_{0,T}(f)\right| \qquad [2.78]$$

	$W_{0,T}(t)$ $(0 \leq t \leq T)$
Rectangular	1
Hamming	$0.54 - 0.46 \cos(2\pi \frac{t}{T})$
Hanning	$0.50 - 0.50 \cos(2\pi \frac{t}{T})$
Blackman	$0.42 - 0.50 \cos(2\pi \frac{t}{T}) + 0.08 \cos(4\pi \frac{t}{T})$
Bartlett	$1 - \left\lvert \frac{2t}{T} - 1 \right\rvert$

Table 2.2. *Continuous time truncation windows*

The first characteristic of a truncation window is the width of the main lobe. This width can be measured by the -3 dB bandwidth, that is B_{-3dB} (T), that is to say the frequency interval for which the Fourier transform module $\left|\hat{w}_{0,N-1}(f)\right|$ is greater than its maximum value $\hat{w}_0$ divided by $\sqrt{2}$ (or, similarly, the interval for which the spectral energy density is greater than half its maximum value). The modulus of the Fourier transform of a window with real values is even. This interval is thus $\left[-\frac{B_{-3dB}(T)}{2}, \frac{B_{-3dB}(T)}{2}\right]$. The equation to be solved to obtain this bandwidth is thus:

$$\left|\hat{w}_{0,T}\left(\frac{B_{-3dB}(T)}{2}\right)\right| = \frac{\hat{w}_{0,T}(0)}{\sqrt{2}} \qquad [2.79]$$

	$\hat{w}_{0,T}(f)$
Rectangular	$e^{-j\pi f T} T \operatorname{sinc}(\pi f T)$
Hamming	$e^{-j\pi f T} T \{0.54 \operatorname{sinc}(\pi f T) + 0.23 \operatorname{sinc}(\pi f T + \pi) + 0.23 \operatorname{sinc}(\pi f T - \pi)\}$
Hanning	$e^{-j\pi f T} T \{0.50 \operatorname{sinc}(\pi f T) + 0.25 \operatorname{sinc}(\pi f T + \pi) + 0.25 \operatorname{sinc}(\pi f T - \pi)\}$
Blackman	$e^{-j\pi f T} T \{0.42 \operatorname{sinc}(\pi f T) + 0.25 \operatorname{sinc}(\pi f T + \pi) + 0.25 \operatorname{sinc}(\pi f T - \pi) + 0.04 \operatorname{sinc}(\pi f T + 2\pi) + 0.04 \operatorname{sinc}(\pi f T - 2\pi)\}$
Bartlett	$e^{-j\pi f T} \frac{T}{2}$

Table 2.3. *Transform of continuous time truncation windows*

That is to say in decibels:

$$W_{\mathrm{dB}}\left(\frac{B_{-3\mathrm{dB}}(T)}{2}\right)=W_{\mathrm{dB}}(0)-3 \qquad [2.80]$$

According to the scale change theorem, a time dependent expansion of a factor $a > 0$ can be translated by a frequential contraction of a factor $\frac{1}{a}$. Thus, we can standardize this bandwidth in the following way:

$$\dot{B}_{-3\mathrm{dB}}=T\,B_{-3\mathrm{dB}}(T) \qquad [2.81]$$

This reduced bandwidth is independent of the amplitude and the length T of the window support, and it depends only on its shape.

Another measurement of the width of the main lobe is the equivalent noise bandwidth $B_{\mathrm{en}}(T)$ of the bandwidth. It is the bandwidth of the window whose spectral energy density is rectangular, having the same integral and same maximum value $\hat{w}_{0,T}^2(0)$ as the density of the window considered (the spectral energy density of a real positive function is always maximum for the zero frequency). We immediately obtain the expression of this bandwidth, in the frequential and time dependent fields (by using Parseval's theorem):

$$B_{\mathrm{en}}(T)=\frac{\int_{-\infty}^{+\infty}\left|\hat{w}_{0,T}(f)\right|^2 df}{\hat{w}_{0,T}^2(0)}$$

$$=\frac{\int_0^T w_{0,T}^2(t)\,dt}{\left(\int_0^T w_{0,T}(t)\,dt\right)^2}$$

Here too we can define the reduced bandwidth independent of T:

$$B_{\mathrm{en}}=T\,B_{\mathrm{en}}(T) \qquad [2.82]$$

The second characteristic of a window is the amplitude of the secondary lobes in relation to the amplitude of the main lobe. A natural measurement is the amplitude of the strongest secondary lobe brought to the amplitude of the main lobe. Consider f_{max} as the abscissa of the maximum of the strongest secondary lobe:

$$f_{max} = \max\left\{ f \neq 0 \middle| \frac{d\left|\hat{w}_{0,T}\right|}{df}(f) = 0 \right\} \quad [2.83]$$

The amplitude of the strongest secondary lobe brought to the amplitude of the main lobe is thus:

$$\frac{\left|\hat{w}_{0,T}\left(f_{max}\right)\right|}{\hat{w}_{0,T}(0)} \quad [2.84]$$

This quantity if independent of *T*. We generally express it in decibels, that is:

$$20\log_{10}\left(\frac{\left|\hat{w}_{0,T}\left(f_{max}\right)\right|}{\hat{w}_{0,T}(0)}\right) = W_{dB}\left(f_{max}\right) - W_{dB}(0) \quad [2.85]$$

Generally, the amplitude of the lobes reduces as the frequency increases. In the plane (log *f*, *WdB* (*f*)), the maxima of the lobes pass through a straight line whose slope can be measured in dB/octave, increase in the value of the ordinate of this line when the frequency is multiplied by 2. This slope measures the "speed" of suppression of the secondary lobes when the frequency increases.

A third characteristic is the positivity of the transform of even windows. In Table 2.3, the Fourier transform of the even window $w_{-\frac{T}{2},\frac{T}{2}}(t)$ is obtained according to the transform of $w_{0,T}(t)$ by multiplying this transform by the $e^{j\pi f t}$ function (according to the delay theorem). We immediately notice that the Bartlett window verifies the positivity property.

Figure 2.7 graphically represents these indices in the case of the rectangular window. Table 2.4 gives the value of these different indices for various windows. Qualitatively speaking, selecting a window representing weak secondary lobes (in order to detect a sine wave of weak power) at the value of a large main lobe (and thus has a loss in resolution: adding the main lobes of two sine waves of similar frequency will give only a single lobe).

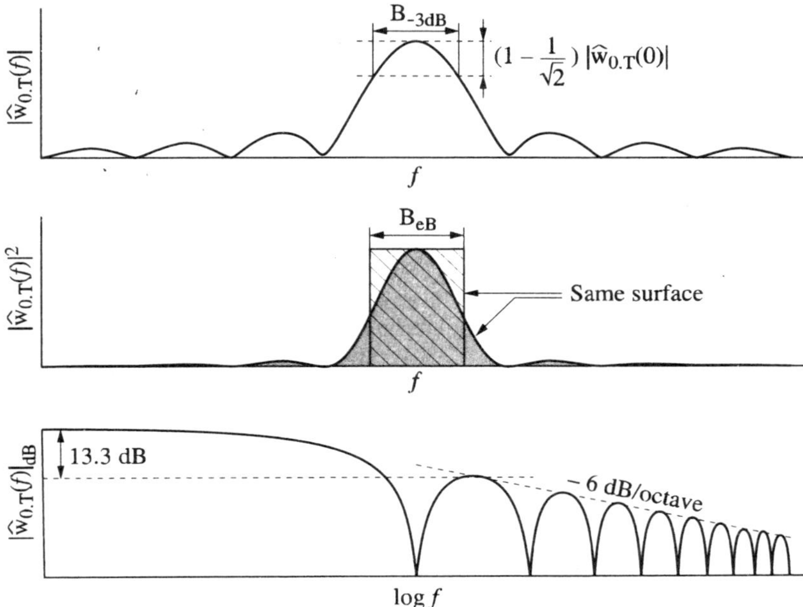

Figure 2.7. *Graphical representation of indices (rectangular window)*

	Band at -3 dB	Band equivalent to noise	Stronger secondary lobes	Decreasing secondary lobes
Rectangular	0.89	1.00	-13.3 dB	-6 dB/octave
Hamming	1.30	1.36	-43.0 dB	-6 dB/octave
Hanning	1.44	1.50	-31.5 dB	-18 dB/octave
Blackman	1.68	1.73	-58.0 dB	-18 dB/octave
Bartlett	1.28	1.33	-26.5 dB	-12 dB/octave

Table 2.4. *Characteristics of continuous time windows*

For digital processing, we record a signal $x(k)$, $k \in Z$ with only a finite number of points N, that is $x(k)$, $0 < k < N-1$. We approximate the Fourier transform $\hat{x}$, i.e.:

$$\hat{x}(v) = \sum_{k=-\infty}^{+\infty} x(k) e^{-j2\pi v k}$$

by a finite sum: $\sum_{k=0}^{N-1} x(k) e^{-j2\pi \nu k}$

This amounts to approximating $\hat{x}$ by the Fourier transform of the signal x multiplied by the gateway $x1_{0,N-1}$ that is $\widehat{x1}_{0,N-1}$. Let us observe the results of this approximation in the case where the signal x is a cisoid of frequency ν_0, of amplitude a and initial phase ϕ:

$$x(k) = a e^{j(2\pi \nu_0 k + \phi)} \qquad \hat{x}(\nu) = a e^{j\phi} \Xi_1(\nu - \nu_0)$$

The product is transformed by the Fourier transform into a circular convolution sum. This convolution with the comb shifted to ν_0 thus shift the Fourier transform of the window to ν_0. We thus obtain:

$$\widehat{x1}_{0,N-1}(\nu) = a e^{j\phi} \hat{1}_{0,N-1}(\nu - \nu_0)$$

Therefore, in the place of the expected pulse at ν_0, we observe Dirichlet's kernel centered on ν_0, and thus a spectral "leakage", as in the case of continuous time signals, is obtained. The discrete time windows designed to correct this phenomenon generally result from the sampling of the continuous time windows (Table 2.5). Figures 2.8 and 2.9 provide the shape of these windows as well as their frequency response in dB, for $N = 32$.

	$\mathbf{w_N(k)\ (0 \le k \le N-1)}$
Rectangular	1
Hamming	$0.54 - 0.46\cos\left(2\pi \frac{\kappa}{N-1}\right)$
Hanning	$0.50 - 0.50\cos\left(2\pi \frac{\kappa}{N-1}\right)$
Blackman	$0.42 - 0.50\cos\left(2\pi \frac{\kappa}{N-1}\right) + 0.08\cos\left(4\pi \frac{\kappa}{N-1}\right)$
Bartlett	$1 - \left\lvert \frac{2\kappa}{N-1} - 1 \right\rvert$

Table 2.5. *Discrete time windows*

We can measure the characteristics of these windows in a manner similar to the case of the continuous windows. The definition of the bandwidth at -3 dB $B_{-3dB}(N)$ becomes:

$$\left|\hat{w}_{0,N-1}\left(\frac{B_{-3\mathrm{dB}}(N)}{2}\right)\right| = \frac{\hat{w}_{0,N-1}(0)}{\sqrt{2}} \qquad [2.86]$$

and its reduced version becomes:

$$\dot{B}_{-3\mathrm{dB}} = N B_{-3\mathrm{dB}}(N) \qquad [2.87]$$

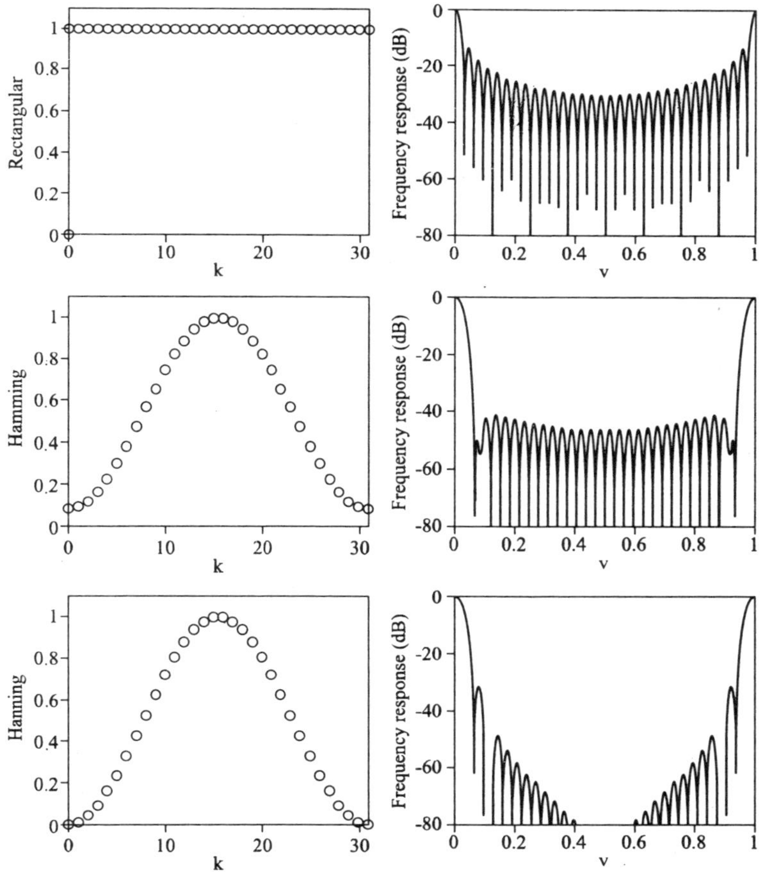

Figure 2.8. *Rectangular, Hamming and Hanning windows and their frequency response (in dB)*

The equivalent noise bandwidth B_{en}, in its reduced form, becomes:

$$\dot{B}_{en} = N \frac{\int_{-\frac{1}{2}}^{+\frac{1}{2}} \left|\hat{w}_{0,N-1}(\nu)\right|^2 d\nu}{\hat{w}_{0,N-1}^2(0)}$$

$$= N \frac{\sum_{k=0}^{N-1} w_{0,T}^2(k)}{\left(\sum_0^{N-1} w_{0,N-1}(k)\right)^2}$$

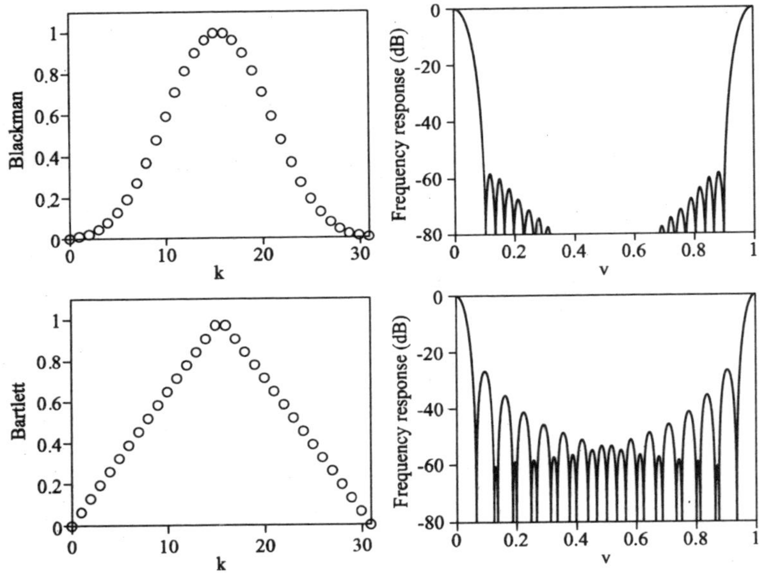

Figure 2.9. *Blackman and Bartlett windows and their frequency response (in dB)*

The amplitude of the secondary lobe is defined in the same way as in continuous. However, these indices that are independent of the length T of the continuous window, depend on N in discrete. By considering the discrete time windows as a sampling to the period $\frac{T}{N}$ of the corresponding continuous time window, we can consider the spectral aliasing phenomenon as negligible for quite large N; the indices defined in discrete thus tend towards their equivalent in continuous time for quite a large N. Thus, Table 2.4 remains valid in discrete time, for quite large values of N (and for lobes sufficiently spaced out of the reduced frequency 1/2 for the decreasing index of secondary lobes).

The definition of discrete time windows often varies in a subtle way, particularly for the windows becoming zero at 0 and $N-1$, such as the Hanning window. We thus lose the information brought by the first and last points. We can correct this oversight by expanding the window to take them into account and increase the resolution by a factor $\frac{N+2}{N}$. For example, the Hanning window is written in some books as:

$$w_{0,N-1}(k) = 0.50 \cos\left(2\pi \frac{k+1}{N+1}\right)$$

2.4. Examples of application

2.4.1. *LTI systems identification*

The Fourier transform is of particular practical interest for the representation of dynamic systems. A system consists of a cause-effect relationship between a batch of input signals and a batch of output signals, indicated by the diagram in Figure 2.10.

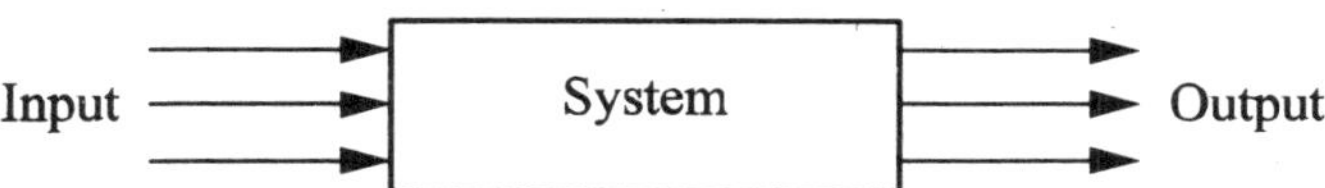

Figure 2.10. *Symbolic representation of a system*

For example, when we take a bath, we use a system with 2 inputs (angular position of hot and cold water taps) and 2 outputs (flow rate and water temperature).

Here, we will limit ourselves to single input and single output systems. Consider a system S of input u and output y; we will provisionally note the cause-effect relation linking the signals y and u in the form:

$$y = S(u)$$

We call the *impulse response* of a system S the Dirac delta function response $S(\delta)$ for continuous time systems, or the Kronecker sequence response $S(\delta)$ for the discrete time systems.

A system S is said to verify the *superposition* principle, or *additivity* principle, if the response to the sum of two inputs is the sum of the responses to each input, whatever the signals u and v may be:

$$S(u + v) = S(u) = S(v)$$

A system S is said to verify the *homogeneity* principle if the responses to two proportional inputs are proportional with the same proportionality coefficient, whatever the signal u and the complex number a may be:

$$S(au) = a\, S(u)$$

A system is said to be *linear* if it verifies the superposition and homogeneity principles.

A system is *time invariant* if a time delay at the input leads to the same time delay at the output. By denoting y_τ as the signal y delayed by τ ($y_\tau(t) = y(t - \tau)$) time, we have, whatever the signals u and y and the real number τ may be:

$$\text{if } y = S(u)$$

$$\text{then } y_\tau = S(u_\tau)$$

A *linear time invariant system* will be indicated with an abbreviated form know as "LTI system".

Numerous physical systems can be considered as linear and invariant over time (particularly, the systems governed by linear differential equations with constant coefficients). The mathematical representation of such systems is well standardized and can be easily interpreted. An LTI system is completely characterized by its pulse response because this is sufficient to calculate the response of the system at any input.

Consider h as the pulse response of the system S:

$$h = S(\delta)$$

Then, the response of the system to an ordinary excitation u is the convolution sum between h and u:

$$S(u) = h \otimes u$$

Let us show this property for a continuous time system, the demonstration in discrete time is analogous, h being the pulse response of the system; we can thus write:

$$t \mapsto h(t) \text{ is the response at } t \mapsto \delta(t)$$

Then, from the time dependent invariance property, for all values of τ:

$$t \mapsto h(t-\tau) \text{ is the response at } t \mapsto \delta(t-\tau)$$

From the homogeneity property, for all values of r and all signals u:

$$t \mapsto u(\tau)h(t-\tau)d\tau \text{ is the response at } t \mapsto u(\tau)\delta(t-\tau)d\tau$$

From the superposition property, we have, for all signals u:

$$t \mapsto \int_{-\infty}^{+\infty} u(\tau)h(t-\tau)d\tau \text{ is the response at } t \mapsto \int_{-\infty}^{+\infty} u(\tau)\delta(t-\tau)d\tau$$

which can be written more concisely by recalling the definition of the convolution sum [2.39] and the formula [2.10]:

$t \mapsto h \otimes u(t)$ is the response at $t \mapsto u(t)$ which completes the demonstration.

Using the Fourier transform, we immediately note that the Fourier transform of the output y of the system is equal to the product of the Fourier transform of the pulse response h and the Fourier transform of the input u.

$$\hat{y} = \hat{h}\,\hat{u} \qquad [2.88]$$

The Fourier transform of the pulse response is known as the transfer function, or harmonic response, or frequency response.

Consider a continuous time LTI system, of pulse response $h(t)$, excited by a cisoid $u(t)$ of amplitude a, frequency f_0 and initial phase ϕ, for all values of t:

$$u(t) = a\,e^{j(2\pi f_0 t+\phi)}$$

Using the Fourier transform, we have, for all values of f:

$$\hat{u}(f) = a e^{j\phi} \delta(f - f_0)$$

Thus, the Fourier transform of the response y is:

$$\begin{aligned}\hat{y}(f) &= \hat{h}(f)\hat{u}(f) \\ &= a\hat{h}(f)e^{j\phi}\delta(f - f_0) \\ &= a\hat{h}(f_0)e^{j\phi}\delta(f - f_0)\end{aligned}$$

Using the inverse Fourier transform, we have, for all values of t:

$$\begin{aligned}y(t) &= a\hat{h}(f_0)e^{j\phi}e^{j2\pi f_0 t} \\ &= a\left|\hat{h}(f_0)\right|e^{j\left[2\pi f_0 t+\phi+\arg\left(\hat{h}(f_0)\right)\right]}\end{aligned}$$

This therefore signifies that the response of an LTI system with pulse response $h(t)$ at a sinusoidal input of frequency f_0 is a sine wave of the same frequency, whose amplitude is multiplied by $\left|\hat{h}(f_0)\right|$ and phase shifted by $\arg\left(\hat{h}(f_0)\right)$. Similar reasoning can be used for the discrete time signals $\left|\hat{h}(f_0)\right|$

The harmonic response is a necessary and sufficient tool for the characterization of LTI systems. It is useful to represent it for real systems (i.e., whose input and output have real values) by one of following three methods:

– the Bode plot, plotted using $\left|\hat{h}(f)\right|$ and $\arg\left(\hat{h}(f)\right)$ according to log(f);

– the Nyquist plot, plotted in the complex plane of $\Im\left(\hat{h}(f)\right)$ according to $\Re\left(\hat{h}(f)\right)$, graduated in function of the frequency;

– the Nichols' plot, plotted in $\left|\hat{h}(f)\right|$ is expressed in dB according to $\arg\left(\left|\hat{h}(f)\right|_{\text{dB}} = 20\log_{10}\left|\hat{h}(f)\right|\right)$ graduated in function of the frequency.

The Nichols and Nyquist plots are of great interest to control engineers, because they help deduce system performance after servo-control, particularly the stability in a closed loop.

These plots can be obtained point by point using a sine wave generator and other adequate measuring instruments (alternating current voltmeter, phase-meter). Another solution consists in calculating from beforehand the Fourier transform of these signals using a recording of the input u and output y of the system and applying the formula for $\hat{u}(f) \neq 0$:

$$\hat{h}(f) = \frac{\hat{y}(f)}{\hat{u}(f)}$$

In practice, these Fourier transforms can be obtained only approximately, by spectral analysis on a sampled recording. The transfer function thus obtained is called "empirical estimator of the transfer function" [LJU 87], which is nothing but the ratio of the discrete Fourier transform of an output signal with the input signal. If there exist frequencies for which $u(f) = 0$, we can use adjustment techniques [IDI 01].

2.4.2. *Monitoring spectral lines*

The Fourier transform, as defined in section 2.2, is a tool suitable for the spectral analysis of sine wave signals. This concept will be discussed in detail in Chapter 5, dedicated to the spectral analysis of random stationary signals. However, the calculation of a Fourier transform of a signal modulated in frequency is of limited interest, because it does not demonstrate the time dependent variation of the spectral content. We present here a preliminary approach to the time-frequency analysis: the sliding Fourier transform helps carry out a spectral analysis close to an instant and whose principle is as follows.

Given a continuous time signal $x(t)$, the sliding Fourier transform according to the time t and the frequency f can be expressed by:

$$R(t,f) = \int_{-\infty}^{+\infty} x(\tau)\, w(\tau - t)\, e^{-j2\pi f \tau} d\tau \qquad [2.89]$$

where $w(t)$ is a finite support even real window. The window $w(\tau - t)$ is thus centered on the analysis instant t. At fixed t_0, the transform $R(t_o, f)$ is nothing but the Fourier transform of the signal $x(t)$ truncated on an interval centered on t_0, and can be considered as a measurement of resemblance between the truncated signal and the sine wave $e^{j2\pi f \tau}$. Inversely, $R(t_0, f)$ can be considered as a measurement of resemblance between the signal $x(t)$ and the sine wave $e^{j2\pi f \tau}$ truncated on the interval centered on t_0

The window $w(t)$ can be rectangular, or smoothed (see section 2.3), to reduce the spectral dilution phenomenon, at the cost of a deterioration in the resolution. We notice here the time-frequency duality principle: monitoring a sliding frequency requires a short support window, to ensure a proper time dependent location, whereas a proper spectral location requires a long support window. The choice of this parameter is thus subjective and depends on the processed signal (slow or fast sliding of the frequency) as well as on the aim in mind (good time dependent or spectral localization).

For a discrete time signal, the sliding Fourier transform can be written as:

$$R(k,\nu) = \sum_{n=-\infty}^{+\infty} x(n) w(n-k) e^{-j2\pi n\nu} \qquad [2.90]$$

As an example, Figure 2.11 represents a signal of statoric current measured on one phase of an induction motor, sampled at 10 ms, as well as the modulus of the corresponding sliding Fourier transform $|R(t,f)|$, calculated using a Hanning window of 40 points.

2.4.3. *Spectral analysis of the coefficient of tide fluctuation*

In France, it is useful to measure the fluctuation m of the tide (difference in height of the water during high tide and low tide) using the coefficient of tide fluctuation c, independent of the geographical point considered and defined by:

$$c = \frac{m}{u} \times 100$$

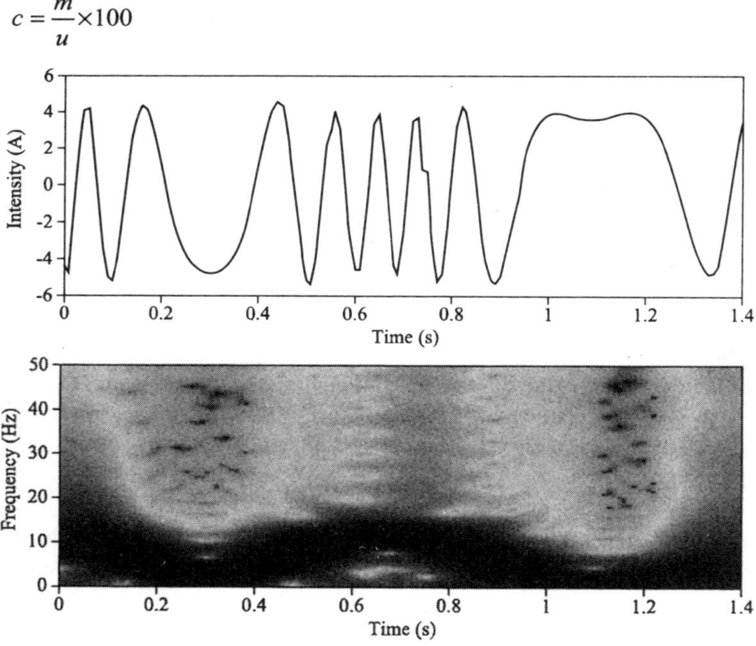

Figure 2.11. *Sliding Fourier transform of statoric current of an asynchronous machine*

where u is a unit of height that depends on the geographical point (for example, 6.10 m at Brest). This coefficient is theoretically between 20 and 120. This coefficient is shown in Figure 2.12 during the year 2001 according to the time of high tides at Mont-Saint-Michel (counted in days starting from 1st January, 0.00 hours), and the corresponding tide fluctuation coefficient.

The duration between two consecutive high tides is not constant. To simplify, between two high tides, we will count the slope of the regression line of the time of high tide according to the high tide index, and we will consider this value as the sampling period for the tide fluctuation coefficient. We obtain:

$$T_c = 12.42 \text{ h} = 0.517 \text{ days}$$

In Figure 2.13 is represented the Fourier transform, calculated using a FFT on 4,096 points, of the weighted tide fluctuation coefficient by a Blackman's window.

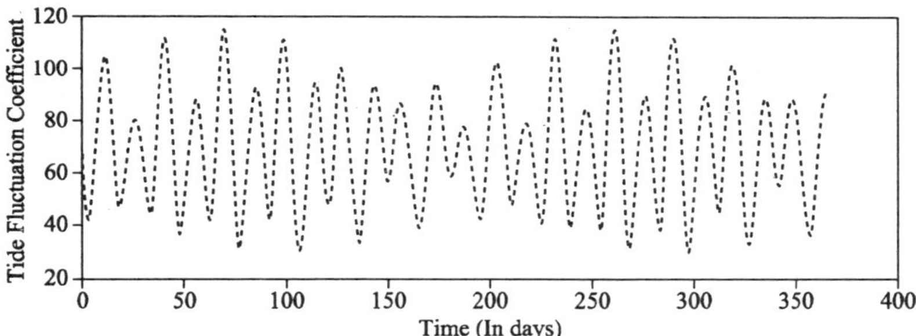

Figure 2.12. *Tide fluctuation coefficient according to time for high tides*

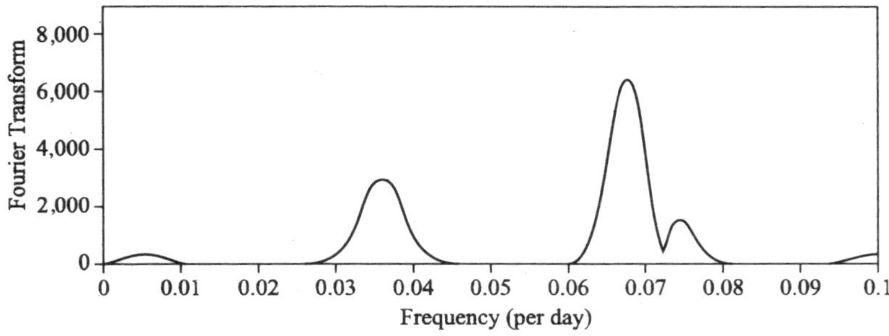

Figure 2.13. *Fourier transform of the weighted tide fluctuation coefficient using a Blackman window*

We clearly see a peak for the frequency 3.63 day^{-1}, i.e. a period of 27.53 days, which can be interpreted as the lunar month. We also see a peak for a frequency which is approximately double, i.e. a period of half a lunar month; in a qualitative way, the spring tides come when we have the following 2 types of alignments: earth-moon-sun, or moon-earth-sun, which explains the half lunar month. We will leave it to the astronomers to interpret this spectral decomposition further.

2.5. Bibliography

[AUG 99] AUGER F., *Introduction à la théorie du signal et de l'information,* Technip, 1999.

[BEL 87] BELLANGER M., *Traitement numérique du signal, théorie et pratique,* Masson, 1987.

[BRI 74] BRIGHAM E., *The Fast Fourier Transform,* Prentice Hall, 1974.

[HAR 78] HARRIS J., "On the use of windows for harmonic analysis with the discrete Fourier transform", *Proceedings of the IEEE,* vol. 66, no. 1, p. 51-84, January 1978.

[IDI 01] ID IER J., Ed., *Approche bayésienne pour les problèmes inverse,* Hermès, 2001.

[KWA 91] KWAKERNAAK H., SIVAN R., *Modem Signals and Systems,* Prentice Hall, 1991.

[LAR 75] DE LARMINAT P., THOMAS Y., *Automatique des systèmes linéaires, Volume 1, Signaux et systèmes,* Dunod, 1975.

[LJU 87] LJUNG L., *System Identification: Theory for the User,* Prentice Hall, 1987.

[MAR 87] MARPLE JR. S., *Digital Spectral Analysis with Applications,* Prentice Hall, 1987.

[MAX 96] MAX J., LACOUME J., *Méthodes et techniques de traitement du signal et applications aux mesures physiques, Volume 1: principes généraux et méthodes classiques.* 5th edition, Masson, 1996.

[OPP 75] OPPENHEIM A., SCHAEFER R., *Digital Signal Processing,* Prentice Hall, 1975.

Chapter 3

Estimation in Spectral Analysis

In this chapter we provide the necessary material for the statistical analysis of various estimators (either of spectrum or of frequencies) described in the body of this book. More precisely, we insist on the methodology of the analysis of estimators, particularly those based on the moments of the signal. After some reminders on estimation theory, we provide results related to the statistical performance of estimators of 1st and 2nd order moments of a stationary random process or a process with a line spectrum. We then show how we can analyze the performance of estimators using these moment estimates.

3.1. Introduction to estimation

3.1.1. *Formalization of the problem*

Before entering into details, we will briefly describe the framework in which we are based and which includes a large part of the problems tackled in the body of this book. The first section is deliberately general. Most of the results will be presented without any demonstration, and anyway the theory of estimation is documented at length in many books; we recommend particularly the following works [AND 71, BRO 91, KAY 93, POR 94, SCH 91, SÔD 89, TRE 71]. We briefly present the elements that help formalize the problem of estimation, define the performance criteria of an estimator and find the optimum estimators. In what follows, we consider using N samples $\{x(n)\}_{n=0}^{N-1}$ of a signal which we group together in a vector; $\boldsymbol{x}_N \in \mathbb{R}^{N\times 1}$:

Chapter written by Olivier BESSON and André FERRARI.

$$x_N = \begin{bmatrix} x(0) \\ x(1) \\ \vdots \\ x(N-1) \end{bmatrix}$$

The vector $\boldsymbol{x}_N$ is considered random; for example, it can come from the sampling of a random process or can be written as the combination of a deterministic signal to which is added a random additive noise. The natural description of $\boldsymbol{x}_N$ is given by the probability density function (PDF), $p(\boldsymbol{x}_N;\ \boldsymbol{\theta})$ which is *parameterized by* $\boldsymbol{\theta} = [\theta_1,\ \theta_2\ \ldots \theta p]^T$, an unknown vector for which we want to estimate the p components. This means that we have a class of different PDF according to the value of $\boldsymbol{\theta}$. Thus it is evident that as $\boldsymbol{\theta}$ influences $p(\boldsymbol{x}_N;\ \boldsymbol{\theta})$, we must be able to infer the value of $\boldsymbol{\theta}$ from $\boldsymbol{x}_N$. Thus we consider an estimator $\hat{\boldsymbol{\theta}}_N$ of $\boldsymbol{\theta}$ from $\boldsymbol{x}_N$, that is to say:

$$\hat{\boldsymbol{\theta}}_N = \boldsymbol{f}(\boldsymbol{x}_N) \qquad [3.1]$$

$\hat{\boldsymbol{\theta}}_N$ being a function of $\boldsymbol{x}_N$ is thus a random vector itself. If the distribution of $\hat{\theta}_N$ provides a complete description of this vector, for estimation we are interested in the following two quantities.

The *bias* represents the *average error* and is given by:

$$\mathrm{b}\left(\hat{\boldsymbol{\theta}}_N\right) = \mathrm{E}\left\{\hat{\boldsymbol{\theta}}_N\right\} - \boldsymbol{\theta} \qquad [3.2]$$

It is desirable to have unbiased estimators, that is to say estimators which provide an exact mean value.

The covariance matrix represents the dispersion around the mean value and its definition is:

$$C\left(\hat{\boldsymbol{\theta}}_N\right) = \mathrm{E}\left\{\left[\hat{\boldsymbol{\theta}}_N - \mathrm{E}\left\{\hat{\boldsymbol{\theta}}_N\right\}\right]\left[\hat{\boldsymbol{\theta}}_N - \mathrm{E}\left\{\hat{\boldsymbol{\theta}}_N\right\}\right]^T\right\} \qquad [3.3]$$

The diagonal elements of the covariance matrix correspond to the respective variances of different elements constituting $\hat{\boldsymbol{\theta}}_N : \boldsymbol{C}(k,k) = \mathrm{var}\left(\hat{\boldsymbol{\theta}}_N(k)\right)$. The terms outside the diagonal give an indication of the degree of correlation between the estimates of different elements. It goes without saying that the smaller the variance, the better the estimation.

REMARK 3.1. We also often consider the root-mean-square error on the estimation of a scalar parameter θ whose definition is given below:

$$\begin{aligned}\text{eqm}\left(\hat{\theta}\right) &= \mathrm{E}\left\{\left(\hat{\theta}-\theta\right)^2\right\} \\ &= \mathrm{b}^2\left(\hat{\theta}\right)+\mathrm{var}\left(\hat{\theta}\right)\end{aligned}$$

3.1.2. *Cramér-Rao bounds*

Ideally, we look for unbiased estimators with minimum variance. This naturally leads to the following question: is there an estimator whose variance is uniformly (i.e. whatever be the value of $\boldsymbol{\theta}$) less than the variance of all other estimators? Before answering this question, we can ask ourselves if a lower bound of the covariance matrix exists; we are thus led to the concept of Cramer-Rao bounds whose definition is given below.

THEOREM 3.1 (CRAMÉR-RAO BOUNDS). *If the PDF $p(\boldsymbol{x}_N;\ \boldsymbol{\theta})$ verifies the regularity condition:*

$$\mathrm{E}\left\{\frac{\partial \ln p\left(\boldsymbol{x}_N;\boldsymbol{\theta}\right)}{\partial \boldsymbol{\theta}}\right\}=0 \quad \forall \theta$$

then, whatever the value of the unbiased estimator $\hat{\theta}_N$ may be, its covariance matrix verifies[1]*:*

$$\boldsymbol{C}\left(\hat{\boldsymbol{\theta}}_N\right)-\boldsymbol{F}_N^{-1}\left(\boldsymbol{\theta}\right)\geq \mathbf{0}$$

where $\boldsymbol{F}_N\left(\boldsymbol{\theta}\right)$ is the Fisher information matrix *given by:*

$$\begin{aligned}\boldsymbol{F}_N\left(\boldsymbol{\theta}\right) &= -\mathrm{E}\left\{\frac{\partial^2 \ln p\left(x_N;\boldsymbol{\theta}\right)}{\partial \boldsymbol{\theta} \partial \boldsymbol{\theta}^T}\right\} \\ &= \mathrm{E}\left\{\frac{\partial \ln p\left(\boldsymbol{x}_N;\boldsymbol{\theta}\right)}{\partial \boldsymbol{\theta}}\frac{\partial \ln p\left(\boldsymbol{x}_N;\boldsymbol{\theta}\right)}{\partial \boldsymbol{\theta}^T}\right\}\end{aligned} \qquad [3.4]$$

---

1 In what follows, the notation $\boldsymbol{A} \geq \boldsymbol{B}$ for two hermitian matrices $\boldsymbol{A}$ and $\boldsymbol{B}$ signifies that $\forall\ \boldsymbol{z}$ the quadratic form $\boldsymbol{z}^H\ (\boldsymbol{A}-\boldsymbol{B})\ \boldsymbol{z} \geq 0$.

and the derivative is evaluated at the true parameter $\boldsymbol{\theta}$*. Furthermore, we can find an estimator which reaches the bound, that is to say* $\boldsymbol{C}(\hat{\boldsymbol{\theta}}) = \boldsymbol{F}_N^{-1}(\boldsymbol{\theta})$ *if and only if:*

$$\frac{\partial \ln p(\boldsymbol{x}_N;\boldsymbol{\theta})}{\partial \boldsymbol{\theta}} = \boldsymbol{F}_N(\boldsymbol{\theta})(\boldsymbol{g}(x)-\boldsymbol{\theta}) \quad [3.5]$$

where $\mathbf{g}$ *is a vector function. This estimator, which is the unbiased estimator with minimum variance, is given by:*

$$\hat{\boldsymbol{\theta}} = \boldsymbol{g}(\boldsymbol{x}) \quad [3.6]$$

and its covariance matrix is then $\boldsymbol{F}_N^{-1}(\boldsymbol{\theta})$. *This estimator is called* efficient.

COROLLARY 3.1. *Suppose we want to estimate* $\boldsymbol{\beta}$ *such that* $\boldsymbol{\beta} = f(\boldsymbol{\theta})$ *from* $\boldsymbol{x}_N$*. Then, we have:*

$$\boldsymbol{C}(\hat{\beta}_N) - \frac{\partial \beta}{\partial \boldsymbol{\theta}^T} \boldsymbol{F}_N^{-1}(\theta) \frac{\partial \beta^T}{\partial \boldsymbol{\theta}} \geq \mathbf{0}$$

where $\partial\beta / \partial\boldsymbol{\theta}^T$ *is the Jacobian matrix whose* (k,ℓ) *element is* $\partial\beta_k / \partial\theta_\ell$. *If* $\hat{\boldsymbol{\theta}}_N$ *is an efficient estimator and if the transformation* $\boldsymbol{f}(\boldsymbol{\theta})$ *is linear, then* $\hat{\beta}_N = \boldsymbol{f}(\hat{\theta}_N)$ *is also efficient. However, if the transformation is non-linear, the efficiency is generally not retained.*

The previous theorem indicates that, under a "generally verified" regularity condition, there exists a lower bound for the covariance matrix. Furthermore, it provides a necessary and sufficient condition for the existence of an efficient estimator (thus with minimum variance). Nevertheless, it is quite rare that this factorization condition is fulfilled and most often we cannot reach the Cramér-Rao bound, except asymptotically, that is to say when the number of observations tends towards infinity.

To illustrate the use of the previous theorem, we now consider the case of *Gaussian signals*. In numerous applications, the signal is distributed according to a normal law:

$$\boldsymbol{x}_N \sim \mathcal{N}(\boldsymbol{s}_N(\boldsymbol{\theta}), \boldsymbol{R}_N(\boldsymbol{\theta}))$$

that is to say:

$$p(\boldsymbol{x}_N;\boldsymbol{\theta})=\frac{1}{(2\pi)^{N/2}\left|\boldsymbol{R}_N(\boldsymbol{\theta})\right|^{1/2}}\exp\left\{-\frac{1}{2}\left[\boldsymbol{x}_N-\boldsymbol{s}_N(\boldsymbol{\theta})\right]^T\boldsymbol{R}_N^{-1}(\boldsymbol{\theta})\left[\boldsymbol{x}_N-\boldsymbol{s}_N(\boldsymbol{\theta})\right]\right\}$$

with $\left|\boldsymbol{R}_N(\boldsymbol{\theta})\right|$ as the determinant of the matrix $\boldsymbol{R}_N(\boldsymbol{\theta})$. $\boldsymbol{s}_N(\boldsymbol{\theta})$ denotes the mean (which can be a useful deterministic signal) and $\boldsymbol{R}_N(\boldsymbol{\theta})$ the covariance matrix. Thus, by applying theorem 3.1, we can obtain the following simplified expression:

$$\begin{aligned}\left|\boldsymbol{F}_N(\boldsymbol{\theta})\right|_{ij}&=\left[\frac{\partial\boldsymbol{s}_N(\boldsymbol{\theta})}{\partial\theta_i}\right]^T\boldsymbol{R}_N^{-1}(\boldsymbol{\theta})\left[\frac{\partial\boldsymbol{s}_N(\boldsymbol{\theta})}{\partial\theta_j}\right]\\&+\frac{1}{2}\mathrm{Tr}\left\{\boldsymbol{R}_N^{-1}(\boldsymbol{\theta})\frac{\partial\boldsymbol{R}_N(\boldsymbol{\theta})}{\partial\theta_i}\boldsymbol{R}_N^{-1}(\boldsymbol{\theta})\frac{\partial\boldsymbol{R}_N(\boldsymbol{\theta})}{\partial\theta_j}\right\}\end{aligned}$$

where $\mathrm{Tr}\{\boldsymbol{A}\}=\sum_{k=1}^{N}\boldsymbol{A}(k,k)$ stands for the trace of the matrix $\boldsymbol{A}$.

We now consider the two following cases.

Case 1

The signal is the sum of a useful deterministic signal $\boldsymbol{s}_N(\boldsymbol{\theta})$ and a noise whose covariance matrix $\boldsymbol{R}_N(\theta)=\boldsymbol{R}_N$ does not depend on $\boldsymbol{\theta}$.

Case 2

The signal is a random process with zero mean [$\boldsymbol{s}_N(\boldsymbol{\theta})\equiv\boldsymbol{0}$] and covariance matrix $\boldsymbol{R}_N(\boldsymbol{\theta})$.

Let us consider the first case. Thus, from the previous formula, we obtain:

$$\boldsymbol{F}_N(\boldsymbol{\theta})=\frac{\partial\boldsymbol{s}_N^T(\boldsymbol{\theta})}{\partial\boldsymbol{\theta}}\boldsymbol{R}_N^{-1}\frac{\partial\boldsymbol{s}_N(\boldsymbol{\theta})}{\partial\boldsymbol{\theta}^T}$$

as the covariance matrix does not depend on $\boldsymbol{\theta}$. Now let us concentrate on the *existence* of an effective estimator. We can write:

$$\frac{\partial\ln p(\boldsymbol{x}_N;\boldsymbol{\theta})}{\partial\boldsymbol{\theta}}=\frac{\partial\boldsymbol{s}_N^T(\boldsymbol{\theta})}{\partial\boldsymbol{\theta}}\boldsymbol{R}_N^{-1}\boldsymbol{x}_N-\frac{\partial\boldsymbol{s}_N^T(\boldsymbol{\theta})}{\partial\boldsymbol{\theta}}\boldsymbol{R}_N^{-1}\boldsymbol{s}_N(\boldsymbol{\theta})$$

We recall that the condition for the existence of an efficient estimator may be written as:

$$\frac{\partial \ln p(\boldsymbol{x}_N;\boldsymbol{\theta})}{\partial \boldsymbol{\theta}} = \boldsymbol{F}_N(\boldsymbol{\theta})\left(\boldsymbol{g}(\boldsymbol{x}_N)-\boldsymbol{\theta}\right) \Rightarrow$$

$$\frac{\partial \boldsymbol{s}_N^T(\boldsymbol{\theta})}{\partial \boldsymbol{\theta}} \boldsymbol{R}_N^{-1} \frac{\partial \boldsymbol{s}_N(\boldsymbol{\theta})}{\partial \boldsymbol{\theta}^T}\boldsymbol{\theta} = \frac{\partial \boldsymbol{s}_N^T(\boldsymbol{\theta})}{\partial \boldsymbol{\theta}} \boldsymbol{R}_N^{-1}\boldsymbol{s}_N(\boldsymbol{\theta}) \Rightarrow$$

$$\frac{\partial \boldsymbol{s}_N(\boldsymbol{\theta})}{\partial \boldsymbol{\theta}^T}\boldsymbol{\theta} = \boldsymbol{s}_N(\boldsymbol{\theta})$$

which is less likely to be true, except if $\boldsymbol{s}_N(\boldsymbol{\theta}) = \boldsymbol{H}_N\boldsymbol{\theta}$. It can be shown ([Hø 00]) that if $\boldsymbol{s}_N(\boldsymbol{\theta})$ is not a linear function (or affine) in $\boldsymbol{\theta}$, there does not exist any efficient estimator, at least with finite N. As an illustration, we consider the following example.

EXAMPLE 3.1. For example, the previous case 1 helps calculate the bound on the parameters of a noisy cisoid:

$$x(n) = A\exp(j2\pi nf+\varphi)+b(n), \boldsymbol{b}_N \sim \mathcal{N}_c\left(\boldsymbol{0},\sigma^2\boldsymbol{I}_N\right)$$

The vector containing the parameters of interest is here[2] $\theta = [\omega \; A \; \varphi]^T$ with $\omega = 2\pi f$. An immediate expansion of the previous results in the case of complex Gaussian signals leads to the following expression of the Fisher matrix:

$$\boldsymbol{F}_N = \frac{1}{\sigma^2}\begin{pmatrix} A^2Q & 0 & A^2P \\ 0 & N & 0 \\ A^2P & 0 & A^2N \end{pmatrix}$$

with $P = \sum_{n=0}^{N-1} n$ and $Q = \sum_{n=0}^{N-1} n^2$. By inverting the matrix $\boldsymbol{F}_N$ we obtain [RIF 74]:

2 In what follows, to simplify the notation, we will sometimes use the angular frequency ω rather than the frequency f, knowing that the two are linked by the relation $\omega = 2\pi f$.

$$\text{var}(\hat{\omega}) \geq \frac{12\sigma^2}{A^2 N\left(N^2-1\right)}$$

$$\text{var}\left(\hat{A}\right) \geq \frac{\sigma^2}{N}$$

$$\text{var}(\hat{\varphi}) \geq \frac{12\sigma^2 Q}{A^2 N^2\left(N^2-1\right)}$$

We note that there does not exist an efficient estimator with finite N in this case; see the previous discussion.

Let us now consider the second case where the mean is $\boldsymbol{s}_N(\boldsymbol{\theta}) = \mathbf{0}$. It thus consists of random signals with zero mean and covariance matrix $\boldsymbol{R}_N(\boldsymbol{\theta})$. The Fisher information matrix is obtained from the derivatives of the covariance matrix in the following manner:

$$\left[\boldsymbol{F}_N(\boldsymbol{\theta})\right]_{ij} = \frac{1}{2}\text{Tr}\left\{\boldsymbol{R}_N^{-1}(\boldsymbol{\theta})\frac{\partial \boldsymbol{R}_N(\boldsymbol{\theta})}{\partial\theta_i}\boldsymbol{R}_N^{-1}(\boldsymbol{\theta})\frac{\partial \boldsymbol{R}_N(\boldsymbol{\theta})}{\partial\theta_j}\right\}$$

While looking for an existence condition of an estimator with minimum variance, we notice rapidly that we end up with implicit equations so that it becomes practically impossible to calculate the estimator with minimum variance. Nevertheless, the expression of the Fisher information matrix sometimes helps attain simple expressions, as shown in the following example.

EXAMPLE 3.2. We try here to calculate the Cramér-Rao bound for an AR process of p order with Gaussian excitation, for which the parameter vector may be written as $\boldsymbol{\theta} = \left[\sigma^2 \; a_1 \; \cdots \; a_p\right]^T$. Friedlander and Porat [FRI 89] demonstrated, using the previous expression, that the Fisher information matrix could be written in this case as:

$$\boldsymbol{F}_N(\boldsymbol{\theta}) = \overline{\boldsymbol{F}}(\boldsymbol{\theta}) + (N-p)\begin{pmatrix} \left(2\sigma^4\right)^{-1} & 0 \\ 0 & \boldsymbol{R}_p \end{pmatrix}$$

where $\overline{\boldsymbol{F}}(\boldsymbol{\theta})$ is a matrix that does not depend on N and where, for $k,\ell = 1, \cdots, p$, $\boldsymbol{R}_p(k,\ell) = \gamma_{xx}(k-\ell)$ with $\gamma_{xx}(.)$ as the correlation function of the process. Thus, $N \to \infty, \bar{\boldsymbol{F}}(\boldsymbol{\theta})$ becomes negligible compared to the second term and we can write:

$$\boldsymbol{C}\left(\hat{\boldsymbol{\theta}}_N\right) \geq \frac{1}{N}\begin{pmatrix} 2\sigma^4 & 0 \\ 0 & \boldsymbol{R}_p^{-1} \end{pmatrix}$$

Before concluding this first part, here is an interesting result by Whittle which expresses the normalized asymptotic Fisher information matrix in certain cases. Let us consider a stationary Gaussian random process of zero mean and power spectral density $S_x(f)$. Then [DZH 86, POR 94]:

$$\begin{aligned} \left[\boldsymbol{F}_0\right]_{k,\ell} &\triangleq \lim_{N\to\infty} N^{-1}\left[\boldsymbol{F}_N(\boldsymbol{\theta})\right]_{k,\ell} \\ &= \frac{1}{2}\int_{-1/2}^{1/2} \frac{1}{S_x^2(f)} \frac{\partial S_x(f)}{\partial \theta_k} \frac{\partial S_x(f)}{\partial \theta_\ell} df \end{aligned} \qquad [3.7]$$

This formula helps us, in particular, obtain extremely simple expressions in the case of ARMA process (see [FRI 84a, FRI 84b] for example).

3.1.3. *Sequence of estimators*

The theoretical elements which have just been given are related to fixed dimensional data vector N. In the context of random processes, we often study the asymptotic behavior of estimators, that is to say when the dimension N of the data vector increases[3]. This gives rise to an estimated sequence $\hat{\theta}_N$ and we study the asymptotic behavior of this sequence, that is to say when $N \to \infty$. Before this we define the types of convergences considered.

Let ξ_N be a sequence of random variables and a_N a series of strictly positive real numbers. We say that ξ_N converges in probability to 0 if, whatever $\delta > 0$ may be:

$$\lim_{N\to\infty} P\left\{\left|\xi_N\right| \geq \delta\right\} = 0$$

3 Another more pragmatic reason is that we rarely know how to carry out a statistical analysis with finite N and most of the results require the hypothesis of a large number of samples [STO 98].

We thus use the notation $\xi_N = o_p(1)$. Similarly, ξ_N is said to be bounded in probability if, whatever $\varepsilon > 0$ may be, there exists $\delta > 0$ such that:

$$P\{|\xi_N| \geq \delta\} < \varepsilon$$

Thus, we note $\xi_N = O_p(1)$. By extension, we consider as convention:

$$\xi_N = o_p(a_N) \Rightarrow a_N^{-1}\xi_N = o_p(1)$$

$$\xi_N = O_p(a_N) \Rightarrow a_N^{-1}\xi_N = O_p(1)$$

When it consists of a sequence of random vectors, the convergence stands for each component of the vector. Thus, we can envisage three types of convergences.

Convergence in probability

We say that ξ_N converges in probability to the random variable ξ if:

$$\xi_N - \xi = o_p(1)$$

Convergence in law

Let $F_{\xi_N}(.)$ be the cumulative distribution function of ξ_N. Then ξ_N converges in law to the random variable ξ if:

$$\lim_{N \to \infty} F_{\xi_N}(x) = F_{\xi}(x)$$

for all values of x for which F_{ξ}, the cumulative distribution function of ξ is continuous.

Mean square convergence

ξ_N converges in mean square towards ξ if:

$$\lim_{N \to \infty} \mathsf{E}\left\{(\xi_N - \xi)^2\right\} = 0$$

RESULT 3.1. *The mean square convergence involves the convergence in probability which in turn involves the convergence in law.*

Let us return to the estimated sequence $\hat{\theta}_N$ We consider two basic criteria, the bias and the consistency.

DEFINITION 3.1. *The estimator* $\hat{\boldsymbol{\theta}}_N$ *is said to be asymptotically unbiased if:*

$$\lim_{N\to\infty} \mathsf{E}\left\{\hat{\boldsymbol{\theta}}_N - \boldsymbol{\theta}\right\} = 0$$

If we can tolerate a bias for a finite number of samples, it is understood that the asymptotic bias must be zero, failing which, even with an infinite number of points of a signal, we cannot find the exact value of the parameter vector. Another important point is the *consistency*.

DEFINITION 3.2. *The estimator* $\hat{\boldsymbol{\theta}}_N$ *is said to be weakly consistent if:*

$$\lim_{N\to\infty} P\left\{\left\|\hat{\boldsymbol{\theta}}_N - \boldsymbol{\theta}\right\| < \delta\right\} = 1 \qquad \forall\delta$$

DEFINITION 3.3. *The estimator* $\hat{\boldsymbol{\theta}}_N$ *is said to be consistent in mean square if:*

$$\lim_{N\to\infty} \mathsf{E}\left\{\left(\hat{\boldsymbol{\theta}}_N - \boldsymbol{\theta}\right)\left(\hat{\boldsymbol{\theta}}_N - \boldsymbol{\theta}\right)^T\right\} = \mathbf{0}$$

The second definition is generally stronger. We refer the reader to [BRO 91, Chapter 6], [POR 94, Chapter 3] for further details on the convergences of estimator sequences and associated properties. One of the key points of a sequence of estimators is the speed with which the estimation errors reduce. These going towards zero, it is natural to standardize $\hat{\boldsymbol{\theta}}_N - \boldsymbol{\theta}$ by an N function such that its order of magnitude is practically independent of N. If there exists an increasing monotonic sequence $d(N)$ such that $d(N)\left(\hat{\boldsymbol{\theta}}_N - \boldsymbol{\theta}\right)$ converges in law towards a random vector $\boldsymbol{\zeta}$, then the distribution of $\boldsymbol{\zeta}$ measures the asymptotic behavior of $\hat{\boldsymbol{\theta}}_N - \boldsymbol{\theta}$[4]. If, for example, $\boldsymbol{\zeta}$ is a random Gaussian vector of zero mean and of covariance matrix $\boldsymbol{\Gamma}$, we can say that $\hat{\boldsymbol{\theta}}_N$ is consistent, asymptotically Gaussian and $\boldsymbol{\Gamma}$ is known as an asymptotically normalized covariance matrix of $\hat{\boldsymbol{\theta}}_N$. However, we must establish a *fine distinction* with:

$$\boldsymbol{\Sigma}(\boldsymbol{\theta}) \triangleq \lim_{N\to\infty} d^2(N)\mathsf{E}\left\{\left(\hat{\boldsymbol{\theta}}_N - \boldsymbol{\theta}\right)\left(\hat{\boldsymbol{\theta}}_N - \boldsymbol{\theta}\right)^T\right\}$$

which is not necessarily equal to $\boldsymbol{\Gamma}$ but which in practice is often the simplest quantity to determine.

---

4 We implicitly consider that all components of $\hat{\boldsymbol{\theta}}_N$ converge at the same speed. If this is not so, we will consider the asymptotic distribution $\boldsymbol{D}(N)\left(\hat{\boldsymbol{\theta}}_N - \boldsymbol{\theta}\right)$ where $\boldsymbol{D}(N) = \operatorname{diag}\left(d_1(N), d_2(N), \ldots, d_p(N)\right)$.

In the same way as we defined the Fisher information matrix with fixed N, we can extend this concept and study the limit:

$$\boldsymbol{F}_0(\boldsymbol{\theta}) \triangleq \lim_{N\to\infty} d^{-2}(N)\boldsymbol{F}_N(\boldsymbol{\theta})$$

which is known as the normalized information matrix. Thus, a consistent estimator $\hat{\boldsymbol{\theta}}_N$, which is asymptotically Gaussian is optimum if $\boldsymbol{\Gamma} = \boldsymbol{F}_0^{-1}(\boldsymbol{\theta})$. Mostly, we speak of an estimator which is asymptotically efficient when:

$$\Sigma(\theta) = \boldsymbol{F}_0^{-1}(\boldsymbol{\theta}) \qquad [3.8]$$

Most of the estimators considered in this book will be mean square convergent with a decrease of the order of $d(N) = \sqrt{N}$, that is to say that their "variance decreases with $1/N$".

REMARK 3.2. The framework selected here for the theoretical analysis of the estimators is the stationary process for which the analysis is generally carried out with a number of points tending towards infinity. It may be noted that this technique is not applicable in numerous cases, for example, for deterministic signals buried in additive noise. For example, if the useful deterministic signal is of finite energy, it goes without saying that the analysis at $N \to \infty$ is not at all valid. Thus, we can resort to an analysis where the signal to noise ratio (SNR) tends towards infinity. Nevertheless, the approach developed in this section will take its meaning, as we always use Taylor series expansions around the mean of the signal (in this case the useful deterministic signal) to carry out analysis, see for example [KAY 93, Chapter 9] or [ERI 93, TRE 85].

3.1.4. *Maximum likelihood estimation*

As we saw in theorem 3.1, the condition for the existence of an efficient estimator is relatively difficult to verify. We then resort to the asymptotic effectiveness. One of the most commonly used means is the maximum likelihood estimator (MLE). The principle is very simple: it consists of finding the parameter $\boldsymbol{\theta}$ which maximizes the likelihood function $p(\boldsymbol{x}_N;\boldsymbol{\theta})$ where $\boldsymbol{x}_N$ corresponds to the observed data. Let us note that as $p(\boldsymbol{x}_N;\boldsymbol{\theta})$ depends on $\boldsymbol{x}_N$, the MLE will be a data function. The MLE looks for the parameter vector $\boldsymbol{\theta}$ which makes the observed data as likely as possible, that is to say:

$$\hat{\boldsymbol{\theta}}_N^{ML} = \arg\max_{\boldsymbol{\theta}} p(\boldsymbol{x}_N;\boldsymbol{\theta})$$

$p(\boldsymbol{x}_N;\boldsymbol{\theta})$, when it is considered as a function of $\boldsymbol{\theta}$, is known as the *likelihood function.* We generally prefer maximizing the logarithm of the likelihood function (log-likelihood[5]) $\mathcal{L}(\boldsymbol{x}_N;\boldsymbol{\theta}) = \ln p(\boldsymbol{x}_N;\boldsymbol{\theta})$.

REMARK 3.3. In the case of a Gaussian process, it is possible to demonstrate that the log-likelihood can be *approximated* by:

$$\mathcal{L}_\omega(\boldsymbol{x}_N;\boldsymbol{\theta}) = -\sum_{j=1}^{(N-1)/2} \ln S_x(f_j,\boldsymbol{\theta}) - \sum_{j=1}^{(N-1)/2} \frac{I(f_j)}{S_x(f_j,\boldsymbol{\theta})} \qquad [3.9]$$

where $I(f_j)$ designates the periodogram of $\boldsymbol{x}_N$ taken at $f_j = j/N$ and $S_x(f_j,\boldsymbol{\theta})$ is power spectral density. This likelihood expression is the discretized version of the likelihood, known as Whittle [WHI62] for which numerous convergence conditions were studied. It is interesting to note that in the case of a process that verifies standard hypotheses, this likelihood can be interpreted as the likelihood associated to the asymptotic distribution of the periodogram which will be derived in section 3.3. The practical importance of this expression in spectral analysis must be highlighted for the numerous cases where only a model of the power spectral density of the signal $S_x(f_j,\boldsymbol{\theta})$ is available.

The ML estimator has the following properties.

THEOREM 3.2 (EFFICIENCY). *If an efficient estimator exists, then the ML estimator produces it.*

THEOREM 3.3 (ASYMPTOTIC PROPERTIES). *If* $p(\boldsymbol{x}_N;\boldsymbol{\theta})$ *is "sufficiently regular", then the ML estimator of* $\boldsymbol{\theta}$ *is consistent, asymptotically Gaussian with normalized covariance matrix* $\boldsymbol{F}_0^{-1}(\boldsymbol{\theta})$.

THEOREM 3.4 (INVARIANCE). *Consider* $\beta = \boldsymbol{f}(\boldsymbol{\theta}) \in \mathbb{R}^q$ *where* $\boldsymbol{f}(\cdot)$ *is a function of* $\mathbb{R}^p$ *in* $\mathbb{R}^q$. *Then, the maximum likelihood estimator of* β *is given by:*

$$\hat{\beta}_N^{ML} = \boldsymbol{f}\left(\hat{\theta}_N^{ML}\right)$$

5 In what follows, the log-likelihood will be given up to an additive constant, the latter does not intervene in the maximization.

If f(.) is not reversible, then $\hat{\beta}_N^{ML}$ is obtained as:

$$\hat{\beta}_N^{ML} = \arg\max_{\beta} \bar{p}(\boldsymbol{x}_N;\beta)$$

$$\bar{p}(\boldsymbol{x}_N;\beta) = \max_{\theta;\beta=\boldsymbol{f}(\boldsymbol{\theta})} p(\boldsymbol{x}_N;\boldsymbol{\theta})$$

Theorem 3.2 is a direct consequence of theorem 3.1; it stipulates that the ML estimator achieves the Cramér-Rao bounds ($\forall$ N) in the case where an efficient estimator definitely exists. For example, this is the case when the signal follows a linear model $\boldsymbol{x}_N = \boldsymbol{H}_N\theta + \boldsymbol{b}_N$ where $\boldsymbol{b}_N$ is a random Gaussian vector of known covariance matrix. Moreover, theorem 3.3 shows that the ML estimator is asymptotically optimum. For example, in the case of $d(N) = \sqrt{N}$ this signifies that:

$$\sqrt{N}\left(\hat{\boldsymbol{\theta}}_N^{ML} - \boldsymbol{\theta}\right) \underset{\text{dist.}}{\longrightarrow} \mathcal{N}\left(\boldsymbol{0}, \boldsymbol{F}_0^{-1}(\boldsymbol{\theta})\right)$$

Finally, theorem 3.4 makes it possible to obtain the maximum likelihood estimator of a function of $\boldsymbol{\theta}$. This theorem can also be applied when it is difficult to find or directly implement $\hat{\beta}_N^{ML}$ but it turns out to be simpler to obtain $\hat{\boldsymbol{\theta}}_N^{ML}$. To sum up, the ML estimator possesses optimality properties for finite N as well as asymptotically. For a wide range of problems, it can be implemented; in particular, it is often systematically used in the case of deterministic signals buried in additive Gaussian noise of known covariance matrix. In the latter case, the ML estimator corresponds to a non-linear least squares estimator. In fact, let us suppose that the signal is distributed as:

$$\boldsymbol{x}_N \sim \mathcal{N}\left(s_N(\boldsymbol{\theta}), \boldsymbol{R}_N\right)$$

with known $\boldsymbol{R}_N$. Then, the log-likelihood may be written as:

$$\mathcal{L}(\boldsymbol{x}_N;\boldsymbol{\theta}) = -\frac{1}{2}\left[\boldsymbol{x}_N - \boldsymbol{s}_N(\boldsymbol{\theta})\right]^T \boldsymbol{R}_N^{-1}\left[\boldsymbol{x}_N - \boldsymbol{s}_N(\boldsymbol{\theta})\right]$$

As a result, the maximum likelihood estimator of $\boldsymbol{\theta}$ takes the following form:

$$\hat{\boldsymbol{\theta}}_N^{ML} = \arg\min_{\boldsymbol{\theta}}\left[\boldsymbol{x}_N - \boldsymbol{s}_N(\boldsymbol{\theta})\right]^T \boldsymbol{R}_N^{-1}\left[\boldsymbol{x}_N - \boldsymbol{s}_N(\boldsymbol{\theta})\right]$$

which corresponds to the best approximation in the least squares sense of the data $\boldsymbol{x}_N$ by the model $\boldsymbol{s}_N(\boldsymbol{\theta})$. In this case we end up with a relatively simple expression. On the contrary, when the covariance matrix is unknown and depends on $\boldsymbol{\theta}$, that is to say when $\boldsymbol{x}_N \sim \mathcal{N}\left(\boldsymbol{s}_N(\boldsymbol{\theta}), \boldsymbol{R}_N(\boldsymbol{\theta})\right)$ the implementation of the ML estimator becomes considerably more complicated, which hampers its use in practice. As an example, for AR or ARMA type linear processes, the exact law is difficult to write (see [SÔD 89, Chapter 7]) which minimizes the use of the ML estimator and attaches greater importance to estimators based on the method of moments. The latter consists of writing $\hat{\boldsymbol{\theta}}_N$ depending on the signal by substituting in the relation expressing the dependence on $\boldsymbol{\theta}$ of the moments of the signal by their natural estimators.

3.2. Estimation of 1st and 2nd order moments

Most of the spectral analysis methods are based on the implicit or explicit use of the 2nd order moments of the signal. This is why the performance of estimators using the moments will significantly depend on the performance of the estimators of moments. At first, we are interested in the estimation of the mean and the covariance function of a *real stationary process.* We consider the case of a random wide sense stationary process whose N samples observed are $\left\{x(n)\right\}_{n=0}^{N-1}$. Results can be obtained with a finite number of points, but the resulting formulae are, contrarily to the asymptotic formulae, generally less explicit. This is why we will often consider the performance of the estimators when $N \to \infty$. At first, we consider stationary signals with a *continuous spectrum* and provide the results for a discrete spectrum in section 3.3. First, some hypotheses are required which are defined now.

HYPOTHESIS 3.1. *The covariances are absolutely summable, that is to say:*

$$\sum_{m=-\infty}^{\infty} \left|c_{xx}(m)\right| \leq C_0 < \infty$$

HYPOTHESIS 3.2. *The 4th order cumulants verify:*

$$\sum_{n=-\infty}^{\infty} \left|\operatorname{cum}(n+k, n, \ell)\right| \leq K_0 < \infty$$

where:

$$\begin{aligned}\operatorname{cum}(k,m,\ell) = \mathrm{E}\left\{x(n)x(n+k)x(n+m)x(n+\ell)\right\} - c_{xx}(k-m)c_{xx}(\ell) \\ -c_{xx}(k-\ell)c_{xx}(m) - c_{xx}(k)c_{xx}(m-\ell)\end{aligned}$$

Let:

$$\begin{aligned}\mu &= \mathsf{E}\{x(n)\} \\ \gamma_{xx}(m) &= \mathsf{E}\{x(n)(x(n+m)\} \\ c_{xx}(m) &= \mathsf{E}\{(x(n)-\mu)(x(n+m)-\mu)\} \\ &= \gamma_{xx}(m)-\mu^2\end{aligned}$$

be the mean, the correlation function and the covariance function of this signal. The usual estimator of the mean is:

$$\hat{\mu} = \frac{1}{N}\sum_{n=0}^{N-1} x(n) \qquad [3.10]$$

This estimator is obviously unbiased:

$$\mathsf{E}\{\hat{\mu}\} = \frac{1}{N}\sum_{n=0}^{N-1}\mathsf{E}\{x(n)\} = \mu$$

We will now study its variance:

$$\begin{aligned}\operatorname{var}(\hat{\mu}) &= \mathsf{E}\{\hat{\mu}^2\} - \mu^2 \\ &= \frac{1}{N^2}\sum_{k=0}^{N-1}\sum_{n=0}^{N-1}\mathsf{E}\{x(k)x(n)\} - \mu^2 \\ &= \frac{1}{N^2}\sum_{k=0}^{N-1}\sum_{n=0}^{N-1} c_{xx}(n-k) \\ &= \frac{1}{N}\sum_{m=-(N-1)}^{N-1}\left(1-\frac{|m|}{N}\right)c_{xx}(m)\end{aligned}$$

The previous formula is exact, whatever the number of points may be. Let us now study the limit of this variance when N tends towards infinity. Using the hypothesis 3.1, we can show that [POR 94, section 4.2]:

$$\lim_{N\to\infty} N\operatorname{var}(\hat{\mu}) = \sum_{m=-\infty}^{\infty} c_{xx}(m) \qquad [3.11]$$

which proves that $\hat{\mu}$ is a consistent in mean square estimator of μ.

REMARK 3.4. The previous calculations establish the mean square consistency of the estimator of the mean and provide an expression of $\lim_{N\to\infty} N\mathrm{var}(\hat{\mu})$ and without making strong hypotheses on the signal. However, if $x(n) - \mu$ is a linear process, that is to say if:

$$x(n)-\mu=\sum_{k=0}^{\infty}\gamma_k e(n-k)$$

where e(n) is a sequence of random independent variables of zero mean and variance σ_e^2, $\sum_{k=-\infty}^{\infty}\left|\gamma_k\right|<\infty$ and $\sum_{k=-\infty}^{\infty}\gamma_k \neq 0$, then [BRO 91, theorem 7.1.2]:

$$\sqrt{N}\left(\hat{\mu}-\mu\right)\overset{as}{\sim}\mathsf{N}\left(0,\sum_{m=-\infty}^{\infty}c_{xx}(m)\right)$$

which means that $\hat{\mu}-\mu$ is asymptotically distributed according to a normal law.

Once the mean has been estimated, we can subtract it from the measurements to obtain a new signal $x(n)-\hat{\mu}$. In what follows, without losing generality, we will suppose that the process has zero mean. The asymptotic results presented later are not affected by this hypothesis [POR 94]. Under these conditions, the two natural estimators of the covariance (or correlation) function are given by:

$$\bar{c}_{xx}(m)=\frac{1}{N-m}\sum_{n=0}^{N-m-1}x(n)x(n+m) \qquad [3.12]$$

$$\hat{c}_{xx}(m)=\frac{1}{N}\sum_{n=0}^{N-m-1}x(n)x(n+m) \qquad [3.13]$$

The mean of each of these estimators is calculated as follows:

$$\mathsf{E}\left\{\bar{c}_{xx}(m)\right\}=\frac{1}{N-m}\sum_{n=0}^{N-m-1}\mathsf{E}\left\{x(n)x(n+m)\right\}=c_{xx}(m) \qquad [3.14]$$

$$\mathsf{E}\left\{\hat{c}_{xx}(m)\right\}=\frac{1}{N}\sum_{n=0}^{N-m-1}\mathsf{E}\left\{x(n)x(n+m)\right\}=\left(1-\frac{m}{N}\right)c_{xx}(m) \qquad [3.15]$$

As a result, $\overline{c}_{xx}(m)$ is an unbiased estimator whereas the bias of $\hat{c}_{xx}(m)$ is $\frac{m}{N}c_{xx}(m)$ However, $\hat{c}_{xx}(m)$ is an asymptotically unbiased estimator of the covariance function. Let us now study the variance and covariances of these estimations and focus on $\hat{c}_{xx}(m)$. We have:

$$\begin{aligned}
\mathrm{E}\left\{\hat{c}_{xx}(m)\hat{c}_{xx}(\ell)\right\} &= \frac{1}{N^2}\sum_{n=0}^{N-m-1}\sum_{k=0}^{N-\ell-1}\mathrm{E}\left\{x(n)x(n+m)x(k)x(k+\ell)\right\} \\
&= \frac{1}{N^2}\sum_{n=0}^{N-m-1}\sum_{k=0}^{N-\ell-1}\left\{c_{xx}(m)c_{xx}(\ell)\right. \\
&\quad + c_{xx}(n-k)c_{xx}(n+m-k-\ell) \\
&\quad + c_{xx}(k-m-n)c_{xx}(k+\ell-n) \\
&\quad \left. + \mathrm{cum}(m,k-n,k+\ell-n)\right\} \\
&= \left(1-\frac{m}{N}\right)\left(1-\frac{\ell}{N}\right)c_{xx}(m)c_{xx}(\ell) \\
&\quad + \frac{1}{N^2}\sum_{n=0}^{N-m-1}\sum_{k=0}^{N-\ell-1}c_{xx}(n-k)c_{xx}(n+m-k-\ell) \\
&\quad + \frac{1}{N^2}\sum_{n=0}^{N-m-1}\sum_{k=0}^{N-\ell-1}c_{xx}(k-m-n)c_{xx}(k+\ell-n) \\
&\quad + \frac{1}{N^2}\sum_{n=0}^{N-m-1}\sum_{k=0}^{N-\ell-1}\mathrm{cum}(m,k-n,k+\ell-n)
\end{aligned}$$

such that:

$$\begin{aligned}
\mathrm{cov}\left(\hat{c}_{xx}(m),\hat{c}_{xx}(\ell)\right) = \frac{1}{N^2}\sum_{n=0}^{N-m-1}\sum_{k=0}^{N-\ell-1} & c_{xx}(n-k)c_{xx}(n+m-k-\ell) \\
& + c_{xx}(k-m-n)c_{xx}(k+\ell-n) \\
& + \mathrm{cum}(m,k-n,k+\ell-n)
\end{aligned} \qquad [3.16]$$

The previous formula uses only the fact that the considered signal is stationary. If, moreover, we set hypothesis 3.2, then we can obtain theorem 3.5 [POR 94, section 4.2].

THEOREM 3.5 (BARTLETT). *Using the absolute summability hypotheses of the covariances and cumulants, we have:*

$$\lim_{N\to\infty} N\operatorname{cov}\left(\hat{c}_{xx}(m),\hat{c}_{xx}(\ell)\right)=\sum_{n=-\infty}^{\infty} c_{xx}(n)c_{xx}(n+m-\ell) + c_{xx}(n+m)c_{xx}(n-\ell) + \operatorname{cum}(n+m,n,\ell) \quad [3.17]$$

The previous theorem is known as Bartlett's formula. We note that if the process is Gaussian, the 4th order cumulants are zero and only the first two terms are retained. Let $\hat{\boldsymbol{c}}=\left[\hat{c}_{xx}(0) \quad \hat{c}_{xx}(1) \quad \ldots \quad \hat{c}_{xx}(M)\right]^T$ be the estimated vector of $\boldsymbol{c}=\left[c_{xx}(0) \quad c_{xx}(1) \quad \ldots \quad c_{xx}(M)\right]^T$ constructed from [3.13].

RESULT 3.2. *If the process is linear,* $\sqrt{N}(\hat{\boldsymbol{c}}-\boldsymbol{c})\overset{as}{\sim}\mathcal{N}(\mathbf{0},\boldsymbol{\Sigma})$ *where the covariance is given by Bartlett's formula; see [BRO 91, p. 228]. This result can also be applied if we suppose that the process is Gaussian [POR 94, p. 105].*

The results given until now concern only random wide sense stationary processes (possibly with linear process hypothesis). Similar results can be demonstrated for other classes of signals, particularly sine wave signals, which are commonly used in spectral analysis. Let us consider the following signal:

$$x(n)=\sum_{k=1}^{p} A_k \sin\left(2\pi n f_k+\phi_k\right)+b(n)=s(n)+b(n) \quad [3.18]$$

where the $\{\phi_k\}$ are assumed to be independent and uniformly distributed over $[0,2\pi[$ and $b(n)$ is a sequence of random variables independent and identically distributed, with zero mean and variance σ^2. The covariance function is then:

$$c_{xx}(m)=\frac{1}{2}\sum_{k=1}^{p} A_k^2\cos\left(2\pi m f_k\right)+\sigma^2\delta(m)=c_{ss}(m)+\sigma^2\delta(m) \quad [3.19]$$

We can thus show that (see for example [STO 89b]):

RESULT 3.3. *If $x(n)$ verifies [3.18],* $\sqrt{N}(\hat{\boldsymbol{c}}-\boldsymbol{c})\overset{as}{\sim}\mathcal{N}(\mathbf{0},\boldsymbol{\Sigma})$ *where the element* (m,ℓ) *of* $\boldsymbol{\Sigma}$ *may be written as:*

$$\Sigma(m,\ell) = 2\sigma^2\left[c_{ss}(\ell+m)+c_{ss}(\ell-m)\right] + \sigma^4\delta(\ell-m)+\left[\mathsf{E}\left\{b^4(n)\right\}-2\sigma^4\right]\delta(\ell)\delta(m) \qquad [3.20]$$

Finally, we would like to mention that very general theoretical results, as they concern signals that are a combination of deterministic and random components (stationary or non-stationary), were published by Dandawaté and Giannakis [DAN 95]. This last article makes it possible to include a very large class of signals and provides analytical expressions of asymptotic distributions of moment estimates. Finally, the reader is asked to read [DAN 95, LAC 97] for the analysis of estimator of higher order moments.

3.3. Periodogram analysis

The periodogram is a natural estimation tool for the power spectral density. Let us recall that its expression is:

$$I(f) = \frac{1}{N}\left|\sum_{n=0}^{N-1} x(n)e^{-j2\pi nf}\right|^2 = \sum_{m=-(N-1)}^{N-1} \hat{\gamma}_{xx}(m)e^{-j2\pi mf} \qquad [3.21]$$

where:

$$\hat{\gamma}_{xx}(m) = \frac{1}{N}\sum_{n=0}^{N-m-1} x(n)x(n+m)$$

designates the biased estimator of the correlation function. The objective of this section is to derive the asymptotic properties of $I(f)$ for a real signal. Firstly we calculate the bias of this estimator. The periodogram mean can directly be written as:

$$\mathsf{E}\{I(f)\} = \sum_{m=-(N-1)}^{N-1}\left(1-\frac{|m|}{N}\right)\gamma_{xx}(m)e^{-j2\pi mf} = \int_{-1/2}^{1/2} S_x(v)W(f-v)\,dv$$

where $W(f) = \frac{1}{N}\left[\frac{\sin(\pi f N)}{\sin(\pi f)}\right]^2$ is the Fourier transform of the triangular window.

As a result, the periodogram is a biased estimator of the power spectral density. Nevertheless, if hypothesis 3.1 is verified, which is always the case for a linear process, then:

$$\lim_{N\to\infty} \mathsf{E}\{I(f)\} = S_x(f) \qquad [3.22]$$

The periodogram is thus an asymptotically unbiased estimator. Moreover, we can establish [POR 94] that:

$$\sum_{m=-\infty}^{\infty} |m||\gamma_{xx}(m)| < \infty \Rightarrow \mathsf{E}\{I(f)\} = S_x(f) + O(1/N)$$

Let us now study the asymptotic distribution of the periodogram. This development is generally done under the hypothesis that the process $x(k)$ is linear [POR 94, PRI 94] or that the cumulants of $x(k)$ are absolutely summable [BRI 81]. We will briefly recall, in the second case, the main steps of the demonstration. We thus move on to hypothesis 3.3.

HYPOTHESIS 3.3. *The cumulants are absolutely summable, that is to say:*

$$\forall k \quad \sum_{u_1,\ldots,u_{k-1}=-\infty}^{+\infty} \left|\mathsf{cum}(u_1,\ldots,u_{k-1})\right| < \infty$$

Let us emphasize that it is possible to define under this hypothesis the multi-spectrum of kth order for all values of k, [LAC 97]:

$$S_x(f_1,\ldots,f_{k-1}) = \sum_{u_1,\ldots,u_{k-1}=-\infty}^{+\infty} \mathsf{cum}(u_1,\ldots,u_{k-1}) e^{-j2\pi\sum_{\ell=1}^{k-1} u_\ell f_\ell}$$

The first step consists of calculating the asymptotic properties of the discrete Fourier transform:

$$d(f) = \frac{1}{\sqrt{N}} \sum_{n=1}^{N} x(n) e^{-j2\pi nf}$$

RESULT 3.4. *Let $k(N)$ be an integer such that $f_k(N) \triangleq k(N)/N \to f_k$ for $k = 1, \ldots, K$ when $N \to \infty$. If $f_k(N) \pm f_\ell(N) \not\equiv 0$ mod(1), then $d(f_k(N))$,*

$k = 1, \ldots, K$ are asymptotically independent and distributed according to $\mathcal{N}(0, S_x(f_k))$.

The approach to demonstrate the previous results is rapidly given below. Thus, it is necessary to calculate the cumulants of the periodogram and its conjugate at different frequencies. Noting that $d(f)^* = d(-f)$, it is thus necessary to calculate $\mathrm{cum}\left(d\left(f_{k_1}\right), \ldots, d\left(f_{k_m}\right)\right)$. By using the multi-linearity of the cumulants, we have:

$$\mathrm{cum}\left(d\left(f_{k_1}\right), \ldots, d\left(f_{k_m}\right)\right) = N^{-\frac{m}{2}} \sum_{n_1, \ldots, n_m = 1}^{N} \mathrm{cum}\left(x(n_1), \ldots, x(n_m)\right) e^{-j2\pi \sum_{q=1}^{m} f_{k_q} n_q} \qquad [3.23]$$

The process $x(n)$ being stationary, we have:

$$\mathrm{cum}\left(x(n_1), \ldots, x(n_m)\right) = \mathrm{cum}\left(n_1 - n_m, \ldots, n_{m-1} - n_m\right)$$

By replacing this expression in [3.23] and by changing the variable $u_k = n_k - n_m$ for $k = 1 \ldots m-1$, we obtain:

$$\mathrm{cum}\left(d\left(f_{k_1}\right), \ldots, d\left(f_{k_m}\right)\right) = N^{-\frac{m}{2}} \sum_{u_1, \ldots, u_{m-1} = -(N-1)}^{N-1} \mathrm{cum}\left(u_1, \ldots, u_{m-1}\right) e^{-j2\pi \sum_{q=1}^{m-1} f_{k_q} u_q} G$$

with:

$$G = \sum_{n_m = 1}^{N} \prod_{e=1}^{m-1} \mathrm{u}\left(u_e + n_m\right) e^{-j2\pi \sum_{q=1}^{m} f_{k_q} n_m}$$

where $\mathrm{u}(n) = 1$ for $1 \leq n \leq N$ and 0 elsewhere. Firstly, we note that:

$$\left| G - \sum_{n_m=1}^{N} e^{-j2\pi \sum_{q=1}^{m} f_{k_q} n_m} \right| \leq \sum_{n_m=1}^{N} \left| \prod_{e=1}^{m-1} \mathrm{u}\left(u_e + n_m\right) - 1 \right| \leq \sum_{e=1}^{m-1} \sum_{n_m=1}^{N} \left| \mathrm{u}\left(u_e + n_m\right) - 1 \right| = \sum_{e=1}^{m-1} \left| u_e \right|$$

We then define:

$$\Delta(f)=\sum_{n_m=1}^{N} e^{-j2\pi f n_m}=e^{-j\pi f(N+1)}\sin(\pi f N)/\sin(\pi f)$$

We can thus write:

$$\begin{aligned}&\Bigg|\text{cum}\left(d\left(f_{k_1}\right),\ldots,d\left(f_{k_m}\right)\right)-\\&\quad N^{-\frac{m}{2}}\sum_{u_1,\ldots,u_{m-1}=-(N-1)}^{N-1}\text{cum}\left(u_1,\ldots,u_{m-1}\right)e^{-j2\pi\sum_{q=1}^{m-1}f_{k_q}u_q}\Delta\left(\sum_{q=1}^{m}f_{k_q}\cdot\right)\Bigg|\\&\quad\le N^{-\frac{m}{2}}\sum_{u_1,\ldots,u_{m-1}=-(N-1)}^{N-1}\left(\sum_{q=1}^{m-1}\left|u_q\right|\right)\left|\text{cum}\left(u_1,\ldots,u_{m-1}\right)\right|\end{aligned}\qquad [3.24]$$

$N^{-\frac{m}{2}}\sum_{q=1}^{m-1}\left|u_q\right|$ tends towards zero when $N\rightarrow\infty$. Application of the dominated convergence theorem with hypothesis 3.3 proves that the difference of the two terms tends towards 0 when $m > 0$. We then obtain:

$$\begin{aligned}\text{cum}\left(d\left(f_{k_1}\right),\ldots,d\left(f_{k_m}\right)\right)&=N^{-\frac{m}{2}}\sum_{u_1,\ldots,u_{m-1}=-(N-1)}^{N-1}\text{cum}\left(u_1,\ldots,u_{m-1}\right)\\&\qquad e^{-j2\pi\sum_{q=1}^{m-1}f_{k_q}u_q}\Delta\left(\sum_{q=1}^{m}f_{k_q}\right)+o(1)\\&=N^{-\frac{m}{2}}\Delta\left(\sum_{q=1}^{m}f_{k_q}\right)S_x\left(f_{k_1},\ldots,f_{k_{m-1}}\right)+o(1)\end{aligned}\qquad [3.25]$$

$m = 1$: we directly verify that $\mathrm{E}\{d(f_{k1}(N))\} = 0$.

$m=2$: if $f_{k1}(N)+f_{k2}(N)\not\equiv 0 \bmod(1), \Delta\left(\sum_{q=1}^{m}f_{k_q}\right)$ is bounded thus:

$$\text{cum}\left(d\left(f_{k1}(N)\right),d\left(f_{k2}(N)\right)\right)\rightarrow 0$$

However, if $f_{k1}(N)+f_{k2}(N)\equiv 0 \bmod(1), N^{-\frac{m}{2}}\Delta\left(\sum_{q=1}^{m} f_{k_q}\right)=1$ thus:

$$\operatorname{cum}\left(d\left(f_{k1}(N)\right), d\left(f_{k2}(N)\right)\right) \rightarrow S_x\left(f_{k1}\right)$$

$m>2$: using a similar reasoning, we can verify that the cumulants of order strictly higher than 2 always tend towards 0.

The cumulants of the Fourier transform tend towards those of a complex process which is jointly Gaussian and independent at the same time. Once the asymptotic distribution of the Fourier transform is obtained, that of the periodogram is obtained directly by noting that:

$$I(f)=\operatorname{Re}\{d(f)\}^2+\operatorname{Im}\{d(f)\}^2$$

RESULT 3.5. *Under the same hypotheses as used previously, the $I(f_k(N))$ are asymptotically independent and distributed according to $S_x(f_k)\chi_2^2/2$ where χ_2^2 designates a chi squared distribution with two degrees of freedom.*

According to properties of the χ_2^2, $\mathrm{E}\{I(f_k(N))\}$ distribution tends towards $S_x(f_k)$ but $\operatorname{var}(I(f_k(N)))$ also tends towards $S_x(f_k)^2$ which shows that *the periodogram is not a consistent estimator of the power spectral density.* However, the independence property of the periodogram bias is the basis of numerous methods generating a consistent estimator of the power spectral density by local averaging of the periodogram [BRI 81].

3.4. Analysis of estimators based on $\hat{c}_{xx}(m)$

The theoretical results of section 3.2 are the basis from which we can analyze the statistical performance of a large number of estimators used in spectral analysis. In fact, the majority of the spectral analysis techniques are based on the use of a set of estimated covariances $\{\hat{c}_{xx}(m)\}_{m=0}^{M}$ or, in other words the estimator may be written as:

$$\hat{\boldsymbol{\theta}}_N=\boldsymbol{g}(\hat{\boldsymbol{c}})$$

We note that the vector $\hat{\boldsymbol{c}}$ considered here consists of estimates of the covariance function. Nevertheless, the approach recommended below can easily be generalized in the case where, for example, $\hat{\boldsymbol{c}}$ contains estimates of moments of higher order, as soon as we know the asymptotic properties of the vector $\hat{\boldsymbol{c}}$. Generally, there are two cases:

1) either the function $\boldsymbol{g}$ is sufficiently simple (linear for example) and we can

relatively easily carry out the analysis of the estimator;

2) or the function $\boldsymbol{g}$ is a more complex function. Nevertheless, it is generally continuous and differentiable (at least close to $\boldsymbol{c}$) and the convergence of $\hat{\boldsymbol{c}}$ towards $\boldsymbol{c}$ allows the use of probabilistic Taylor series expansions.

Before describing the approach in detail, some results related to the asymptotic theory may be recalled; see [BRO 91, Chapter 6] and [POR 94, Appendix C].

RESULT 3.6. *Let* ξ_N *be a sequence of random vectors of dimension* p *such that* $\xi_N - \xi = O_p(a_N)$ and $\lim_{N\to\infty} a_N = 0$. *If g is a function of* $\mathbb{R}^p$ in $\mathbb{R}$ *such that the derivatives* $\partial g/\partial \xi$ *are continuous in a neighborhood of* ξ, *then:*

$$g(\xi_N) = g(\xi) + \left[\frac{\partial g}{\partial \xi}\right]^T (\xi_N - \xi) + o_p(a_N)$$

RESULT 3.7. *Consider* ξ_N and ζ_N *as two sequences of random vectors of dimension p such that* $\xi_N - \zeta = o_p(1)$ and ξ_N *converges in law towards* ξ. *Then,* ζ_N *also converges in law towards the random vector* ξ. Using the two previous results, we can then demonstrate (see [BRO 91, Chapter 6]) the following result.

RESULT 3.8. *Let* ξ_N *be a sequence of random vectors of dimension* p *such that* $a_N^{-1}(\xi_N - \xi) \overset{as}{\sim} \mathcal{N}(\mathbf{0}, \Sigma)$ and $\lim_{N\to\infty} a_N = 0$ *Consider:*

$$\boldsymbol{g}(\boldsymbol{x}) = \begin{bmatrix} g_1(\boldsymbol{x}) & g_2(\boldsymbol{x}) & \dots & g_q(\boldsymbol{x}) \end{bmatrix}^T$$

a function of $\mathbb{R}^p$ *in* $\mathbb{R}$ *continuous and differentiable in a neighborhood of* ξ. *We note by* $\boldsymbol{D} = \partial \boldsymbol{g} / \partial \boldsymbol{x}^T$ *the matrix* $q \times p$ *of partial derivatives, evaluated at* ξ *and we suppose that* $\boldsymbol{D}\Sigma\boldsymbol{D}^T$ *has all its diagonal elements as non-zero. Then:*

$$a_N^{-1}(\boldsymbol{g}(\xi_N) - \boldsymbol{g}(\xi)) \overset{as}{\sim} \mathcal{N}\left(\mathbf{0}, \boldsymbol{D}\Sigma\boldsymbol{D}^T\right)$$

The previous results apply directly to the analysis of estimators of the power spectral density. In fact, many parametric spectral analysis techniques consist of estimating a vector of parameters $\boldsymbol{\theta}$ and using $\hat{\boldsymbol{\theta}}_N$ in the expression of the power spectral density $S_x(f, \boldsymbol{\theta}) : \hat{S}_x(f, \boldsymbol{\theta})_N \triangleq S_x\left(f, \hat{\boldsymbol{\theta}}_N\right)$. If $\hat{\boldsymbol{\theta}}_N$ is an estimator of $\boldsymbol{\theta}$ such that $a_N^{-1}\left(\hat{\boldsymbol{\theta}}_N-\boldsymbol{\theta}\right)\overset{as}{\sim}\mathcal{N}(\mathbf{0},\boldsymbol{\Sigma})$ and $\lim_{N\to\infty}a_N=0$ then the result 3.8 helps us immediately conclude that:

$$a_N^{-1}\left(\hat{S}_x(f,\theta)_N-S_x(f,\theta)\right)\overset{as}{\sim}\mathcal{N}\left(\mathbf{0},\boldsymbol{d}^T(f,\theta)\boldsymbol{\Sigma}\boldsymbol{d}(f,\theta)\right)$$

where $\boldsymbol{d}(f,\boldsymbol{\theta})=\partial S_x(f,\boldsymbol{\theta})/\partial\boldsymbol{\theta}$. Using similar reasoning and by basing ourselves on the corollary 3.1, Friedlander and Porat suggest in [FRI 84b] the use of:

$$\boldsymbol{d}^T(f,\boldsymbol{\theta})\boldsymbol{F}_N^{-1}(\boldsymbol{\theta})\boldsymbol{d}(f,\boldsymbol{\theta})$$

where the Fisher matrix is obtained from Whittle's formula [3.7], as the lower bound to the variance of $\hat{S}_x(f,\boldsymbol{\theta})_N$:

$$\operatorname{var}\left(\hat{S}_x(f,\boldsymbol{\theta})_N\right)\geq\frac{2}{N}\boldsymbol{d}^T(f,\boldsymbol{\theta})\left\{\int_{-1/2}^{1/2}\frac{\boldsymbol{d}(f,\boldsymbol{\theta})\boldsymbol{d}(f,\boldsymbol{\theta})^T}{S_x^2(f,\boldsymbol{\theta})}df\right\}^{-1}\boldsymbol{d}(f,\boldsymbol{\theta})$$

Results 3.6, 3.7 and 3.8 also help analyze the performance of a large class of estimators based on $\hat{c}$. The aim of this section is not to be exhaustive but rather to focus on an analytical approach; we will illustrate the latter on two specific cases:

1) estimation of parameters of an AR model;

2) estimation of the frequency of a noisy cisoid by subspace methods.

3.4.1. *Estimation of parameters of an AR model*

We thus analyze the estimation of parameters of an autoregressive model and the estimation of the frequencies of the poles of this model. Let us suppose that $x(k)$ is an autoregressive process of order ρ whose power spectral density may be written as:

$$S_x(f)=\frac{\sigma^2}{A(f)A^*(f)}$$

$$A(f)=1+\sum_{k=1}^{p}a_ke^{-j2\pi kf}$$

Let us study the estimation of the parameter vector $\boldsymbol{a} = a_1\ a_2\ \ldots\ a_p]^T$. As shown in Chapter 6, the natural estimator of $\boldsymbol{a}$ may be written as:

$$\hat{\boldsymbol{a}} = -\widehat{\boldsymbol{R}_p}^{-1}\hat{\boldsymbol{r}} \qquad [3.26]$$

$$\widehat{\boldsymbol{R}_p} = \begin{pmatrix} \hat{c}_{xx}(0) & \hat{c}_{xx}(-1) & \cdots & \cdots & \hat{c}_{xx}(-p+1) \\ \hat{c}_{xx}(1) & \hat{c}_{xx}(0) & \cdots & \cdots & \hat{c}_{xx}(-p+2) \\ \vdots & \vdots & \ddots & & \vdots \\ \vdots & \vdots & & \ddots & \vdots \\ \hat{c}_{xx}(p-1) & \hat{c}_{xx}(p-2) & \cdots & \cdots & \hat{c}_{xx}(0) \end{pmatrix}$$

$$\hat{\boldsymbol{r}} = \begin{pmatrix} \hat{c}_{xx}(1) \\ \hat{c}_{xx}(2) \\ \vdots \\ \hat{c}_{xx}(p) \end{pmatrix}$$

Thus, from results 3.8 and 3.2, it directly follows that:

$$\sqrt{N}(\hat{\boldsymbol{a}} - \boldsymbol{a}) \overset{as}{\sim} \mathcal{N}\left(\boldsymbol{0}, \frac{\partial \boldsymbol{a}}{\partial \boldsymbol{c}^T}\Sigma\frac{\partial \boldsymbol{a}^T}{\partial \boldsymbol{c}}\right)$$

The matrix of derivatives $\partial\boldsymbol{a}/\partial\boldsymbol{c}^T$ can easily be calculated by using the Yule-Walker equations [POR 94]. The previous approach directly used result 3.8, which is very general. Nevertheless, the approach may be followed more specifically for each case. For the AR estimation, we can write:

$$\begin{aligned} \sqrt{N}(\hat{\boldsymbol{a}} - \boldsymbol{a}) &= -\sqrt{N}\widehat{\boldsymbol{R}_p}^{-1}\left(\hat{r} + \widehat{\boldsymbol{R}_p}\boldsymbol{a}\right) \\ &= -\widehat{\boldsymbol{R}_p}^{-1}\left\{\sqrt{N}\left[(\hat{r} - r) + \left(\widehat{\boldsymbol{R}_p} - R_p\right)\boldsymbol{a}\right]\right\} \\ &= -\boldsymbol{R}_p^{-1}\left\{\sqrt{N}\left[(\hat{r} - r) + \left(\widehat{\boldsymbol{R}_p} - R_p\right)\boldsymbol{a}\right]\right\} + o_p(1) \\ &= -\boldsymbol{R}_p^{-1}\sqrt{N}\varepsilon + o_p(1) \end{aligned} \qquad [3.27]$$

The vector $\sqrt{N}\varepsilon$ is asymptotically distributed according to a normal law of $\boldsymbol{0}$ mean and covariance matrix $\boldsymbol{S}$ whose element (k,ℓ) is:

$$\boldsymbol{S}(k,\ell) = \sum_{i=0}^{p}\sum_{j=0}^{p} a_i a_j \sigma_{k-i,\ell-j}$$

with $\sigma_{k.\ell} = \lim_{N\to\infty} N \operatorname{cov}\left(\hat{c}_{xx}(k), \hat{c}_{xx}(\ell)\right)$ whose expression is given in theorem 3.5. Thus, from equation [3.27] and using result 3.7, we can conclude that:

$$\sqrt{N}\left(\hat{\boldsymbol{a}}-\boldsymbol{a}\right) \overset{as}{\sim} \mathcal{N}\left(\mathbf{0}, \boldsymbol{R}_p^{-1}\boldsymbol{S}\boldsymbol{R}_p^{-1}\right)$$

In the case where the process is regarded as Gaussian, by using the Yule-Walker equations for an AR process, we can demonstrate that:

$$\boldsymbol{S}(k,\ell) = \sigma^2 c_{xx}(k-\ell)$$

which means that $\mathbf{S} = \sigma^2 \boldsymbol{R}_p$ and finally:

$$\sqrt{N}\left(\hat{\boldsymbol{a}}-\boldsymbol{a}\right) \overset{as}{\sim} \mathcal{N}\left(\mathbf{0}, \sigma^2 \boldsymbol{R}_p^{-1}\right)$$

Example 3.2 shows that the asymptotically normalized Fisher information matrix is $\sigma^2 \boldsymbol{R}_p^{-1}$ which implies that the estimator $\hat{\boldsymbol{a}}$ is asymptotically optimum for an AR Gaussian process.

Let us now study the estimation of frequencies of poles of the model. In order to simplify, we consider $p = 2q$ as even and note that:

$$\begin{aligned} A(z) &= \sum_{k=0}^{p} a_k z^{-k} = \prod_{k=1}^{p}\left(1 - z_k z^{-1}\right) \\ &= \prod_{k=1}^{q}\left(1-\rho_k e^{j\omega_k} z^{-1}\right)\left(1-\rho_k e^{j\omega_k} z^{-1}\right) \end{aligned}$$

We will note $\hat{z}_k = \hat{\rho}_k e^{j\hat{\omega}_k}$ as the estimates of poles obtained from $\hat{A}(z) = 1 + \sum_{k=1}^{p} \hat{a}_k z^{-k}$ We attempt to determine the asymptotic distribution of $\sqrt{N}\left(\hat{\boldsymbol{\omega}}-\boldsymbol{\omega}\right)$ where $\hat{\boldsymbol{\omega}} = \left[\hat{\omega}_1 \cdots \hat{\omega}_q\right]^T$ and $\boldsymbol{\omega} = \left[\omega_1 \cdots \omega_q\right]^T$ For this, the approach is the same as the one that links the statistics of $\sqrt{N}\left(\hat{\boldsymbol{a}}-\boldsymbol{a}\right)$ to those of $\sqrt{N}\left(\hat{\boldsymbol{c}}-\boldsymbol{c}\right)$. The functional $\hat{\boldsymbol{\omega}} = \boldsymbol{g}\left(\hat{\boldsymbol{a}}\right)$ being continuous and differentiable, we attempt to perform a Taylor series expansion that helps obtain the asymptotic distribution of $\sqrt{N}\left(\hat{\boldsymbol{\omega}}-\boldsymbol{\omega}\right)$. See [STO 89a] for the calculation and details of theoretical elements; we simply indicate that we have:

$$\sqrt{N}\left(\hat{\omega}-\omega\right) = \boldsymbol{F}\boldsymbol{G}\left\{\sqrt{N}\left(\hat{\boldsymbol{a}}-\boldsymbol{a}\right)\right\} + o_p(1)$$

with:

$$F = \left(\begin{array}{ccc|ccc} \frac{\beta_1}{\alpha_1^2+\beta_1^2} & & 0 & \frac{-\alpha_1}{\alpha_1^2+\beta_1^2} & & 0 \\ & \ddots & & & \ddots & \\ 0 & & \frac{\beta_q}{\alpha_q^2+\beta_q^2} & & & \frac{-\alpha_q}{\alpha_q^2+\beta_q^2} \end{array} \right)$$

$$G = \begin{pmatrix} \boldsymbol{h}_1^T \\ \vdots \\ \boldsymbol{h}_q^T \\ \boldsymbol{g}_1^T \\ \vdots \\ \boldsymbol{g}_q^T \end{pmatrix}$$

and, for $k = 1, \cdots, q$:

$$\begin{aligned} \alpha_k &= \left[\rho_k^{-2} \cos(\omega_k) \quad \cdots \quad p\rho_k^{-p-1} \cos(p\omega_k) \right] \boldsymbol{a} \\ \beta_k &= \left[\rho_k^{-2} \sin(\omega_k) \quad \cdots \quad p\rho_k^{-p-1} \sin(p\omega_k) \right] \boldsymbol{a} \\ \boldsymbol{g}_k &= \left[\rho_k^{-2} \sin(\omega_k) \quad \cdots \quad \rho_k^{-p-1} \sin(p\omega_k) \right]^T \\ \boldsymbol{h}_k &= \left[\rho_k^{-2} \cos(\omega_k) \quad \cdots \quad \rho_k^{-p-1} \cos(p\omega_k) \right]^T \end{aligned}$$

As a result, we obtain the following asymptotic distribution:

$$\sqrt{N}(\hat{\omega} - \omega) \overset{as}{\sim} \mathcal{N}\left(\boldsymbol{0}, \boldsymbol{FGR}^{-1}\boldsymbol{SR}^{-1}\boldsymbol{G}^T\boldsymbol{F}^T\right)$$

3.4.2. *Estimation of a noisy cisoid by MUSIC*

Some estimators are obtained by the minimization of a function, that is to say:

$$\hat{\boldsymbol{\theta}}_N = \arg\min_{\boldsymbol{\theta}} J_N(\boldsymbol{\theta})$$

where the criteria $J_N(\boldsymbol{\theta})$ depends either on the data or on the correlation function. In general, the approach followed consists of expanding the derivative of the criteria into a Taylor series and using the asymptotic arguments presented earlier. This is the case in the MUSIC algorithm presented in Chapter 8 for which the frequencies are determined as arguments which minimize the square norm of the projection of $\boldsymbol{a}(\omega)$ on the estimated noise subspace $\hat{\boldsymbol{V}}_b$:

$$J_N(\omega)=\frac{1}{2}\left|\widehat{\Pi}_b^H\,\boldsymbol{a}(\omega)\right|^2,\widehat{\Pi}_b=\widehat{\boldsymbol{V}}_b\widehat{\boldsymbol{V}}_b^H \tag{3.28}$$

We intend to give the main steps of the methodology of the analysis in the mono-component case:

$$x(n)=Ae^{j\omega_0 n}+b(n)$$

For a complete expansion we can refer to [STO 91], for example. Let $\hat{\omega}$ be the frequency that minimizes the criteria [3.28]. The convergence of $\hat{\omega}$ towards ω_0 helps use a Taylor expansion of $J'_N(\hat{\omega})$ around ω_0:

$$0=J'_N(\omega_0)+J''_N(\omega_0)(\hat{\omega}-\omega_0)+\cdots$$

Limiting ourselves to the first order, the error made asymptotically on the estimation of ω_0 is thus:

$$\hat{\omega}-\omega_0\simeq -J'_N(\omega_0)/J''_N(\omega_0) \tag{3.29}$$

with:

$$J'_N(\omega_0)=\operatorname{Re}\left\{\boldsymbol{a}(\omega)^H\,\widehat{\Pi}_b\boldsymbol{d}(\omega)\right\}$$

$$J''_N(\omega_0)=\operatorname{Re}\left\{\boldsymbol{a}(\omega)^H\,\widehat{\Pi}_b\boldsymbol{d}'(\omega)+\boldsymbol{d}(\omega)^H\,\widehat{\Pi}_b\boldsymbol{d}(\omega)\right\}$$

Asymptotically, the projection $\widehat{\Pi}_b$ converges towards Π_b which allows us to write:

$$\begin{aligned}J''_N(\omega_0)&\simeq\operatorname{Re}\left\{\boldsymbol{a}(\omega)^H\,\Pi_b\boldsymbol{d}'(\omega)+\boldsymbol{d}(\omega)^H\,\Pi_b\boldsymbol{d}(\omega)\right\}\\&=\operatorname{Re}\left\{\boldsymbol{d}(\omega)^H\,\Pi_b\boldsymbol{d}(\omega)\right\}=2\operatorname{Re}\left\{\boldsymbol{d}(\omega)^H\left(\boldsymbol{I}-\boldsymbol{V}_s\boldsymbol{V}_s^H\right)\boldsymbol{d}(\omega)\right\}\\&=M\left(M^2-1\right)/12\end{aligned} \tag{3.30}$$

The calculation of $J'_N(\omega_0)$ can be done using analysis of the perturbation on projection operators presented in [KRI 96]:

$$\widehat{\Pi}_b = \Pi_b + \delta\Pi_b + \cdots$$

By limiting ourselves to the first order, we thus obtain:

$$J'_N(\omega_0) \simeq \text{Re}\left\{\boldsymbol{a}(\omega)^H \delta\Pi_b \boldsymbol{d}(\omega)\right\} \qquad [3.31]$$

with:

$$\delta\Pi_b = -\Pi_s\left(\hat{\boldsymbol{R}} - \boldsymbol{R}\right)\boldsymbol{S}^{\#} - \boldsymbol{S}^{\#}\Delta\left(\hat{\boldsymbol{R}} - \boldsymbol{R}\right)\Pi_s, \boldsymbol{S} = |\boldsymbol{A}|^2 \boldsymbol{a}(\omega)\boldsymbol{a}(\omega)^H$$

where $\boldsymbol{S}^{\#}$ designates the pseudo-inverse of $\boldsymbol{S}$. By substituting [3.31] and [3.30] in [3.29] we thus obtain an expression of the error on $\hat{\omega}$ according to $\hat{\boldsymbol{R}} - \boldsymbol{R}$ which helps link the variance of $\hat{\omega}$ to the variance of $\hat{\boldsymbol{R}}$. Then, the asymptotic arguments (for example. results 3.6, 3.7 and 3.8) help complete the calculation of the asymptotic variance of $\hat{\omega}$ (see [STO 91] for a detailed analysis).

3.5. Conclusion

In this chapter, we studied the methods of analysis of estimators commonly used in spectral analysis. The approach consists most often of linking the estimation errors on the parameters $\boldsymbol{\theta}$ to the estimation errors on the moments of the signal and use the asymptotic arguments based on Taylor expansions. If this approach does not cover all the cases envisaged, it provides answers to a large number of problems of spectral analysis. See the books mentioned in the bibliography for information on how to tackle other possible cases.

3.6. Bibliography

[AND 71] ANDERSON T., *The Statistical Analysis of Time Series,* John Wiley, New York, 1971.

[BRI 81] BRILLINGER D., *Time Series: Data Analysis and Theory,* McGraw-Hill, 1981.

[BRO 91] BROCKWELL P. J., DAVIS R. A., *Time Series: Theory and Methods,* Springer Series in Statistics, Springer Verlag, Berlin, 2nd edition, 1991.

[DAN 95] DANDAWATÉ A., GLANNAKIS G., "Asymptotic theory of kth-order cyclic moment and cumulant statistics", *IEEE Transactions Information Theory,* vol. 41, no. 1, p. 216-232, January 1995.

[DZH 86] DZHAPARIDZE K., *Parameter Estimation and Hypothesis Testing in Spectral Analysis of Stationary Time Series,* Springer Verlag, New York, 1986.

[ERI 93] ERIKSSON A., STOICA P., SODERSTRÔM T., "Second-Order Properties of MUSIC and ESPRIT Estimates of Sinusoidal Frequencies in High SNR Scenarios", *Proceedings IEE-F Radar, Sonar Navigation,* vol. 140, no. 4, p. 266-272, August 1993.

[FRI 84a] FRIEDLANDER B., "On the Computation of the Cramer-Rao Bound for ARMA Parameter Estimation", *IEEE Transactions Acoustics Speech Signal Processing,* vol. 32, no. 4, p. 721-727, August 1984.

[FRI 84b] FRIEDLANDER B., PORAT B., "A General Lower Bound for Parametric Spectrum Estimation", *IEEE Transactions Acoustics Speech Signal Processing,* vol. 32, no. 4, p. 728-733, August 1984.

[FRI 89] FRIEDLANDER B., PORAT B., "The Exact Cramer-Rao Bound for Gaussian Auto-regressive Processes", *IEEE Transactions Aerospace and Electronic Systems,* vol. 25, no. 1, p. 3-8, January 1989.

[HO 00] HOST-MADSEN A., "On the Existence of Efficient Estimators", *IEEE Transactions Signal Processing,* vol. 48, no. 11, p. 3028-3031, November 2000.

[KAY 93] KAY S., *Fundamentals of Statistical Signal Processing: Estimation Theory,* Prentice Hall, Englewood Cliffs, NJ, 1993.

[KRI 96] KRIM H., FORSTER P., "Projections on Unstructured Subspaces", *IEEE Transactions Signal Processing,* vol. 44, no. 10, p. 2634-2637, October 1996.

[LAC 97] LACOUME J.-L., AMBLARD P.-O., COMON P., *Statistics of Greater Order for Signal Processing,* Masson, Paris, 1997.

[POR 94] PORAT B., *Digital Processing of Random Signals: Theory & Methods,* Prentice Hall, Englewood Cliffs, NJ, 1994.

[PRI 94] PRIESTLEY M. B., *Spectral Analysis and Time Series,* Monographs and Textbooks, Academic Press, 8th edition, 1994.

[RIF 74] RIFE D., BOORSTYN R., "Single-tone Parameter Estimation from Discrete-time Observations", *IEEE Transactions Information Theory,* vol. 20, no. 5, p. 591-598, September 1974.

[SCH 91] SCHARF L., *Statistical Signal Processing: Detection, Estimation and Time Series Analysis,* Addison Wesley, Reading, MA, 1991.

[SOD 89] SÔDERSTRÔM T., STOICA P., *System Identification,* Prentice Hall International, London, UK, 1989.

[STO 89a] STOICA P., SÔDERSTRÔM T., TI R., "Asymptotic Properties of the High-Order Yule-Walker Estimates of Sinusoidal Frequencies", *IEEE Transactions Acoustics Speech Signal Processing,* vol. 37, no. 11, p. 1721-1734, November 1989.

[STO 89b] STOICA P., SÔDERSTRÔM T., TI R, "Overdetermined Yule-Walker Estimation of the Frequencies of Multiple Sinusoids: Accuracy Aspects", *Signal Processing,* vol. 16, no. 2, p. 155-174, February 1989.

[STO 91] STOICA P., SÔDERSTRÔM T., "Statistical Analysis of MUSIC and Subspace Rotation Estimates of Sinusoidal Frequencies", *IEEE Transactions Signal Processing,* vol. 39, no. 8, p. 1836-1847, August 1991.

[STO 98] STOICA P., "Asymptomania?" *IEEE Signal Processing Magazine,* vol. 15, no. 1, p. 16, January 1998.

[TRE 71] TREES H. V., *Detection, Estimation and Modulation Theory,* John Wiley, New York, 1971.

[TRE 85] TRETTER S., "Estimating the Frequency of a Noisy Sinusoid by Linear Regression", *IEEE Transactions Information Theory,* vol. 31, no. 6, p. 832-835, November 1985.

[WHI 62] WHITTLE P., "Gaussian Estimation in Stationary Time Series", *Bulletin of the International Statistical Institute,* vol. 39, p. 105-129, 1962.

Chapter 4

Time-Series Models

4.1. Introduction

The time series $x(k)$, during the processing of a signal, generally comes from the periodic sampling of a continuous time signal $x(t)$, that is to say:

$$x(k) \triangleq x(t)\rfloor_{t=k.T_c}$$

The main idea of the models of such a series is to suggest a dynamic equation, which takes into account the time dependence variation of $x(k)$ most often in terms of its past $\{x(k-1), x(k-2)\ldots, x(-\infty)\}$. This equation can take into account any future values of $m > k$, but the processing real time constraints can be limited to causal dynamic equations:

$$x(k) = F\left[x(k-1), x(k-2)\ldots, x(-\infty)\right] \qquad [4.1]$$

The operator F [.] can be of a varied nature – linear, non-linear, of finite or non-finite dimension, etc.

The practical use of such dynamic equations is to obtain compact expression (thus F [.] in finite dimension) equipped with a set of parameters θ, which help adapt this model $F\left[x(k-1), x(k-2)\ldots, x(-\infty), \theta\right]$ has a certain variety of each actual signal. The effectiveness of a model will be measured on the one hand in a range of

---

Chapter written by Francis CASTANIÉ.

signal classes of which it can model the variation with considerable accuracy, and on the other hand on the number of characteristics of the signal which it can model.

If equation [4.1] represents no physical reality (which will practically always be the case), the adjustment of the model to a physical signal can only be done in the case of a precise criterion of accuracy applied to a characteristic of the signal, which is also precise and limited. We will see later in this paragraph that if a model is fitted to respect the characteristics at the M order of the signal (moments for example), it will not adjust itself to characteristics of a different order or laws, etc.

Thus, the choice of a model is very important and never easy: it affects the actual success of methods, which result from its use.

As part of the current book, the model must also provide a relation between the parameters θ and the spectral properties to be measured. If, for example, we are interested in the power spectral density (PSD) $S_x(f)$ of $x(t)$, the model retained $F\left[x(k-1),x(k-2)\ldots,x(-\infty),\theta\right]$ must satisfy the relation:

$$\theta \Rightarrow S_x(f,\theta) \qquad [4.2]$$

All estimators $\hat{\theta}$ of θ can thus provide a corresponding estimator of $S_x(f)$ from a relation of the type:

$$\hat{S}_x(f) \triangleq \hat{S}_x\left(f,\hat{\theta}\right) \qquad [4.3]$$

This approach must be verified for all the spectral characteristics of interest: stationary or non-stationary PSD and ESD, multispectra, etc.

The fundamental problem that such an approach poses in the context of spectral analysis is to evaluate how the estimation errors on $\hat{\theta}$ will affect $\hat{S}_x(f)$. The tools present in section 3.4 make it possible to tackle it as a particular case of an estimator function, and by a limited expansion to evaluate the bias, variance, covariances, etc. of $\hat{S}_x(f)$ from the equivalent characteristics of $\hat{\theta}$. The application to the model known as *autoregressive* given in section 3.4.1 is a paradigm of this type of problem.

From among the models, the most classical ones are strongly linked to "system" representation of the signal: the signal observed $x(k)$ is conceptually represented as the output of a system, fed by an input $u(k)$. We select as neutral an input as possible, which is described by a restricted number of parameters, the largest number of parameters possible being carried by the transfer function such as the one represented in Figure 4.1.

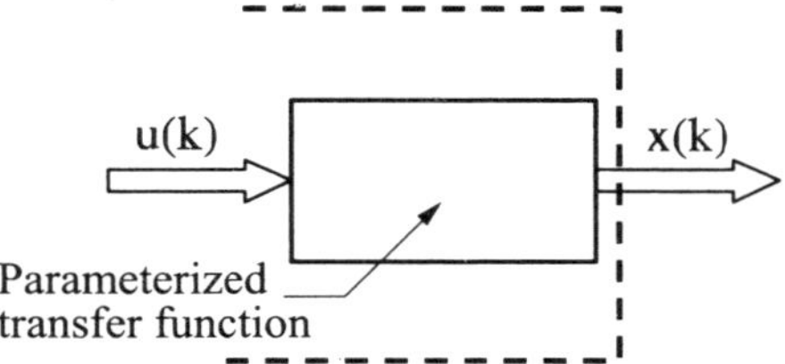

Figure 4.1. *Modeling "system" approach*

This approach was historically decisive for signal processing, then for images: all identification tools, essentially developed by the control engineer, and to a lesser degree in numerical analysis where it can be used to solve the main problem, which is that of estimation of parameters of the model which help adjust $x(k)$ to the physical signal. The difference (and additional difficulty) with the problem of identification of the systems is that the virtual input $u(k)$ is not physically accessible. Thus we will have to base ourselves only on the observation of $x(k)$ to build the estimators. The hypotheses, which we will make on the transfer function of the system, will help classify the models: we will discuss the linear or non-linear, stationary or non-stationary models, etc., according to the corresponding properties of the system. As for the inputs, they will be selected for their "neutrality", that is to say they would limit the class $x(k)$ generated as much as possible: the most current would be the null mean white noise, but we will also find excitations derived from a series of periodic pulses (such as voice models), from a finite set of pulses (multi-pulse models), or even a single pulse, etc.

4.2. Linear models

When the system appearing in Figure 4.1 is linear, the very large set of results on this class of systems can be used: this is a very strong motive to accept the limitation to this particular case and we will discover that it will provide a class of very efficient models. The system could be invariant by translation or otherwise. In the first case, it could lead to models of stationary signals, if $u(k)$ itself is stationary; in the second case, we would have access to non-stationary models otherwise known as "evolutionary"

4.2.1. *Stationary linear models*

If we suppose that the system appearing in Figure 4.1 is invariant and linear, with an impulse response $\{h(k)\}$, and thus of transmittance in z:

$$\breve{h}(z) \triangleq ZT(h(k)) \qquad [4.4]$$

we know that the input-output relation may be written as:

$$x(k) = u \otimes h(k) = \sum_{m=-\infty}^{+\infty} u(m) h(m-k) \qquad [4.5]$$

which may be reduced to:

$$x(k) = \sum_{m=-\infty}^{k} u(m) h(m-k) = \sum_{m=-0}^{+\infty} h(m) u(m-k) \qquad [4.6]$$

if we limit ourselves to the causal models.

The system is rigorously defined by an infinity of parameters $\{h(k)\}$, except if we further reduce to finite impulse response systems (FIR) (see [BEL 89, NAJ 02]). This limited case known as MA models will be tackled later.

To help have a more general system, but which is however defined by a *finite* number of parameters, to this day only one solution is known to us: the system is described by a linear recurrence of finite order, with constant coefficients:

$$x(k) = -\sum_{n=1}^{p} a_n x(k-n) + \sum_{n=0}^{q} b_n u(k-n) \qquad [4.7]$$

This corresponds to a transmittance, which is a rational fraction:

$$h(z) = \frac{\sum_{n=0}^{q} b_n z^{-n}}{\sum_{n=0}^{p} a_n z^{-n}} \qquad a_0 = 1 \qquad [4.8]$$

The parameter vector θ_h of the system is in this case:

$$\theta_h^T = \left| a_1 \cdots a_p ; b_0 \cdots b_q \right|$$

with $\dim(\theta_h) = p + q + 1$.

This consists of an infinite impulse response filter (IIR), according to the usual system theory terminology. The link between this parameter vector of finite length the impulse response of infinite length $\{h(k)\}$ appearing in equation [4.5], is visible when we expand into a series:

$$\breve{h}(z) = \frac{\sum_{n=0}^{q} b_n z^{-n}}{\sum_{n=0}^{p} a_n z^{-n}} = \sum_{n=0}^{+\infty} h(n) z^{-n} \qquad [4.9]$$

which will not be reduced to a polynomial except for a trivial case when the numerator is divisible by the denominator.

The particular case of FIR mentioned earlier is only the one where $a_n = 0$ for $n \neq 0$. In this case, the transmittance can be written in the form of a polynomial:

$$\breve{h}(z) = \sum_{n=0}^{q} b_n z^{-n} = \sum_{n=0}^{q} h(n) z^{-n} \qquad [4.10]$$

We note that if the input $u(k)$ is considered inaccessible (it is a virtual input), but some of its probabilistic characteristics are known, we can perhaps calculate some probabilistic characteristics of $x(k)$, considering that we know how the said characteristics propagate through an invariant linear system. The simplest case is when we are only interested in the 2nd order characteristics (Wiener-Lee relations; see [KAY 88, MAR 87]), which may be written in discrete time as:

$$\begin{aligned} S_x(z) &= S_u(z)\breve{h}(z)\breve{h}^*\left(\frac{1}{z^*}\right) \text{ for } z \in \mathbb{C} \\ S_x\left(e^{i2\pi v}\right) &= S_u\left(e^{i2\pi v}\right)\left|\breve{h}\left(e^{i2\pi v}\right)\right|^2 \text{ for } z = e^{i2\pi v} \end{aligned} \qquad [4.11]$$

It is no doubt compelling to take for $S_u(z)$ the simplest PSD possible:

$$S_u(z) = C^{te} = \sigma^2$$

that is to say for $u(k) = n(k)$ a *zero mean white noise*. In this case, the PSD can be written as:

$$S_x\left(e^{i2\pi v}\right) = \sigma^2 \left|\breve{h}\left(e^{i2\pi v}\right)\right|^2 \qquad [4.12]$$

which we can write to demonstrate the role of the parameter vector:

$$S_x\left(e^{i2\pi\nu},\theta_h\right)=\sigma^2\left|\breve{h}\left(e^{i2\pi\nu},\theta_h\right)\right|^2 \qquad [4.13]$$

The complete parameter vector, including the single scalar parameter of the input is thus:

$$\theta=\left|\sigma^2\ \theta_h^T\right|^T=\left|\sigma^2\ a_1\cdots a_p\ b_0\cdots b_q\right|^T \qquad [4.14]$$

This type of model of a recursive system of finite order excited by a stationary and null mean white noise is strictly speaking (see [BOX 70]), a model known as ARMA (for AutoRegressive with Adjusted Mean), noted as ARMA(p,q). However, we will find the name ARMA more widely accepted in literature, once the system complies with equation [4.7] for an input, which is not necessarily white noise. The case of the FIR system is noted as MA (for moving average), the recursion terms (or autoregressive) having disappeared.

We can clearly see in equation [4.13] the role that the free parameters θ play: they help conform to PSD of $x(k)$, in the particular framework of rational spectra (in fact rational trigonometric spectra). It is also clear that if we imagine an estimation method of θ, which helps adjust the PSD of a model to that of a physical signal (which will be dealt with in Chapter 6), this will not imply anything (in all generality) as for the adjustment of other characteristics, such as higher statistics, laws, etc.

4.2.2. *Properties*

4.2.2.1. *Stationarity*

We can show ([BOX 70, KAY 88]) that the stationarity of $x(k)$ is only linked to the stability of the linear system (it is understood that the excitation $\eta(k)$ is itself stationary): all stable systems will lead to a stationary model. Thus, the stationary condition does not imply that the parameters $\{a_n\}$, *via* the decomposition of the denominator of:

$$\breve{h}(z)=\frac{\sum_{n=0}^{q}b_n z^{-n}}{\sum_{n=0}^{p}a_n z^{-n}}=\frac{B(z)}{A(z)}$$

in the polar form:

$$\{a_n\} \Rightarrow A(z) = \sum_{n=0}^{p} a_n z^{-n} = \prod_{n=1}^{p} \left(1 - p_n z^{-1}\right) \Rightarrow \{p_n\}$$

and the stability condition:

$$|p_n| < 1 \forall n \qquad [4.15]$$

4.2.2.2. *Moments and spectra*

If we limit ourselves to zero mean signals, the general theorems given in Chapter 1 on the multispectra make it possible to obtain the general relation [NIK 93]:

$$S_{xx\cdots x}(\nu_1,\ldots,\nu_{n-1}) = S_{nn\cdots n}(\nu_1,\ldots,\nu_{n-1}).\, H(\nu_1,\ldots,\nu_{n-1}) \qquad [4.16]$$

with:

$$H(\nu_1,\ldots,\nu_{n-1}) = \breve{h}\left(\exp(i2\pi\nu_1)\right)\ldots\breve{h}\left(\exp(i2\pi\nu_{n-1})\right)\breve{h}^*\left(\exp\left(i2\pi[\nu_1+\ldots+\nu_{n-1}]\right)\right)$$

If we consider the whiteness of the excitation $n(k)$ at the considered order:

$$S_{nn\ldots n}(\nu_1,\ldots,\nu_{n-1}) = C_{nn\ldots n}(0,\ldots,0)\,\forall \nu_1,\ldots,\nu_{n-1}$$

we obtain an expression, which generalizes equation [4.12] to all orders. This helps obtain useful spectral relations for this book. However, the equation for generating the signal $x(k)$ of equation [4.7] helps write a direct fundamental relation on the moments.

To start with, let us consider the autocorrelation function of $x(k)$, $\gamma_{xx}(m)$. It is easy to directly show (for example, see [KAY 88]) from equation [4.7] that:

$$\begin{aligned}\gamma_{xx}(m) &= E\left(x(k)x^*(k-m)\right)\\ &= \sum_{n=1}^{p} a_n\,\gamma_{xx}(m-n) + \sigma^2 \sum_{n=m}^{q} b_n h^*(m-n) \qquad 0 \le m \le q\\ &= \sum_{n=1}^{p} a_n\,\gamma_{xx}(m-n) \qquad m \ge q+1\end{aligned} \qquad [4.17]$$

The last line of this equation has many applications: it shows that if we consider the correlation function $\gamma_{xx}(m)$ only for m ≥ q + 1, the recurrence (identical to the recursive part of the equation of the signal itself) establishes a very simple link between the correlation and the AR $\{a_k\}$ parameters. In the form of a matrix, for m = K, K +1, …, $K + p - 1$ ($\forall K \geq q + 1$) it may be written as:

$$\mathbf{Ra} = -\mathbf{r} \qquad [4.18]$$

with:

$$\mathbf{R} = \begin{vmatrix} \gamma_{xx}(K-1) & \cdots & \gamma_{xx}(K-p) \\ \gamma_{xx}(K) & \cdots & \gamma_{xx}(K-p+1) \\ \vdots & \ddots & \vdots \\ \gamma_{xx}(K+p-2) & \cdots & \gamma_{xx}(K-1) \end{vmatrix}$$

and:

$$\mathbf{r} = \begin{vmatrix} \gamma_{xx}(K) \\ \gamma_{xx}(K+1) \\ \vdots \\ \gamma_{xx}(K+p-1) \end{vmatrix}$$

This allows us to know a if we know a sufficient interval of $\gamma_{xx}(m)$, by inverting equation [4.18] under regularity conditions of **R**: it will be the Yule-Walker relation, which is the key to most of the estimators with AR parameters (see section 6.2):

$$\mathbf{a} = -\mathbf{R}^{-1}\mathbf{r} \qquad [4.19]$$

We will note that the matrix to be inverted has a very particular structure known as Toeplitz:

$$[\mathbf{R}]_{ij} = \gamma_{xx}(K-1+i-j)$$

and, as a result, rapid inversion algorithms exist (see [BOX 70, KAY 88]).

System of linear equations, rapid algorithms: all the components were assembled in this equation to create a historical success!

The relations between the MA $\{b_n\}$ parameters and the correlation function cannot be directly used from equation [4.17]: this relation shows a combination between these parameters and the coefficients of the impulse response, which is in itself a non-linear combination of $\{b_n\}$ and $\{a_n\}$. Section 6.2 suggests methods for resolutions, which are essentially non-linear.

In the case of the pure AR(p) model, which it particularly important for the applications (that is to say with $b_0 = 1$, $b_1 = \ldots = b_q = 0$), we can select $K = 1$ in the Yule-Walker equations: the matrix $\mathbf{R}$ is the most symmetrical.

4.2.2.3. *Relation with Wold's decomposition*

For example, it is shown in [LAC 00], that *all* 2nd order *stationary random processes*, with some conditions of regularity, may be written in the form:

$$x(k) = \sum_{n=0}^{\infty} c_n \xi(k-n) + \upsilon(k) \qquad [4.20]$$

where the series of $\{\xi(n)\}$ is a series of non-correlated variables (noted as *innovations*), the series of coefficients $\{c_n\}$ is determined by projection techniques and the term "singular" $\upsilon(k)$ has the special feature of being exactly predictable from its past.

This extremely general decomposition is Wold's decomposition.

The singular term $\upsilon(k)$ represents a *deterministic* signal (it is one of the definitions, amongst many, of the determinism of signals). If we exclude it from the model, the rest of the decomposition is similar to equation [4.5]: the series $\{\xi(n)\}$ is a white noise series, and the coefficients $\{c_n\}$ play the role of pulse response coefficients. Can we say that any stationary random signal (which has a Wold's decomposition) is an ARMA signal?

This is often found in literature, but is wrong: if we limit equation [4.20] in the case of $\upsilon(k) = 0\ \forall k$, we obtain a common convolution equation:

$$x(k) = c \otimes \xi(k)$$

and if ZT $\breve{\xi}(z)$ of $\xi(k)$ exists, it plays the role of a transmittance. The fundamental difference with the transmittance $\breve{h}(z)$ defining the ARMA model is that $\breve{\xi}(z)$ *is not,* generally, *a rational fraction.*

There are many examples of non-rational PSD signals in physics: spectra of turbulence signals, signals with long dependence and a fractal nature, etc. (see [ABR 02]).

However, we see that the ARMA models will be powerful approximants of stationary signals, at least in their spectral characteristics, because the theory of rational approximation states that we can always approximate a function $\breve{\xi}(z)$ as closely as we want, under relatively large convergence constraints, by a rational fraction of sufficient p and q degrees. This supports the common practical approach, which is to approximate the PSD of a physical signal by a series of the ARMA(p,q) models of increasing orders, until some fidelity criterion on the PSD of the model is satisfied (see Chapter 6).

4.2.3. *Non-stationary linear models*

The general way of writing a non-stationary linear model is as follows:

$$x(k) = \sum_{l} h(l,k)u(l) \tag{4.21}$$

where the pulse response $h(l, k)$ is not reduced to a kernel with 1 dimension $h(l-k)$ as in the translation invariant case.

If we reduce the class of the models we are interested in to non-correlated excitations ($u(l) = n(l)$ is a white noise of power σ^2), we obtain a simple relation on the correlation functions:

$$\begin{aligned} R_{xx}(k,m) &= E\left[x(k)x^*(m)\right] = \sigma^2\left\{\sum_{l} h(l,k)h^*(l,m)\right\} \\ &= \sigma^2 H(k,m) = \sigma^2 \breve{H}(k,k-m) \end{aligned}$$

The first question is to decide a definition of an "*evolutive spectrum*", because, as mentioned in section 1.2.2, the beautiful uniqueness of the definition of PSD does not exist outside the stationary framework: Chapter 9 of the current book gives an introduction to this problem and a complete study can be found in another book of the same treatise ([HLA 05]). Whatever this definition may be, it is based on a two-dimensional Fourier transform of $H(k, m)$ or on one of its versions, such that $\tilde{H}(k,k-m)$.

The non-stationary linear models must therefore present systems with pulse responses having two variables, equipped with a vector of parameter θ, which helps, as in the stationary case, to adjust them to the spectral characteristics of physical signals.

There are several approaches to destationarize the system, which is at the heart of the model:

1) We can select an unstable system: to destabilize the system is to generate a non-stationary model.

2) We can use inputs that are non-stationary.

3) The parametric system in itself can be selected as *varying*, that is to say, it is no more limited to being a translation system.

And of course the combination of the three possibilities.

The first solution has limited possibilities: in fact, to destabilize the model, we cannot push one or more poles outside or on the unit circle.

A pole outside the unit circle results in an exponential instability, which is very "explosive" for most applications. The poles which are exactly on the unit circle can lead to purely periodic or polynomial instabilities: they are the root of the ARIMA models which are introduced in [BOX 70] (see Chapter 6 and [HLA 05]).

The non-stationary inputs change the problem: how can this non-stationarity be easily described and parameterized? The *multi-pulse* models are one of the pragmatic approaches to this problem (see a brief introduction in section 9.3.2).

A more attractive "system" approach consists of using linear filters with time variable parameters: in equation [4.7], the parameters $\{a_n,b_n\}$ become time functions k. What follows is a possible way of writing:

$$x(k)=-\sum_{n=1}^{p}a_n(k-n)x(k-n)+\sum_{n=0}^{q}b_n(k-n)u(k-n) \qquad [4.22]$$

If the excitation $u(k)$ is a centered stationary white noise $n(k)$, some properties of the ARMA models of the previous paragraph can be generalized, which will be presented in section 6.1. These models are known as *evolutionary ARMA*.

A main problem still needs to be resolved: the parameterization. The time-varying filter and all its characteristics, such as the impulse response, are known if we know function sets $\{a_n(k), b_n(k)\}$. Thus, a suitable parameterization of these families needs to be found, which can be either deterministic or random. If we limit ourselves to the case of deterministic functions, the standard approach consists of projecting them on a finite set of functions. If we select a set of base functions $\{f_m(k)\}$, the projection may be written as:

$$a_n(k) = \sum_{m=1}^{M} a_{nm} f_m(k)$$ [4.23]

$$b_n(k) = \sum_{m=1}^{M} b_{nm} f_m(k)$$

and the global parameter vector of the model is now:

$$\theta = \left[a_{11} a_{12} \cdots a_{1M} a_{21} a_{22} \cdots a_{2M} \cdots a_{pM} \right]^T$$ [4.24]

One of the key properties of these models is the extension of the Yule-Walker theorem, applied to the following vectors:

$$\mathbf{x}(k) = \begin{bmatrix} f_1(k)x(k) \\ \vdots \\ f_M(k)x(k) \end{bmatrix} \text{ and } \mathbf{u}(k) = \begin{bmatrix} f_1(k)u(k) \\ \vdots \\ f_M(k)u(k) \end{bmatrix}$$

and:

$$\mathbf{a}_n = \begin{bmatrix} a_{n0} \\ \vdots \\ a_{nM} \end{bmatrix} \text{ and } \mathbf{b}_n = \begin{bmatrix} b_{n0} \\ \vdots \\ b_{nM} \end{bmatrix}$$

which leads to the following way of writing a dynamic model:

$$x(k) = -\sum_{n=1}^{p} \mathbf{a}_n^T \mathbf{x}(k-n) + \sum_{n=0}^{q} \mathbf{b}_n^T \mathbf{u}(k-n)$$ [4.25]

This equation helps write an expression of the Yule-Walker type:

$$E\left[\begin{pmatrix} \mathrm{x}(k-1) \\ \vdots \\ \mathrm{x}(k-p) \end{pmatrix} \left(\mathrm{x}^T(k-1) \cdots \mathrm{x}^T(k-p)\right)\right]\theta$$
$$= -E\left[\begin{pmatrix} \mathrm{x}(k-1) \\ \vdots \\ \mathrm{x}(k-p) \end{pmatrix} x(k)\right]$$

which is also a basic tool with which we define estimators of parameter vector θ from equation [4.24].

4.3. Exponential models

A class of models with special focus on signal processing: these are the complex exponential base decompositions.

The main reason is the property that these functions are eigenfunctions of invariant linear systems. The exponential models are historically created by Prony (1795), and were unearthed and revived in the 1970s by Kumaresan and Tufts [KUM 82] (see also [KAY 88, MAR 87]). They can be broken down into deterministic, noisy deterministic and random models.

4.3.1. *Deterministic model*

Supposing we can write a (deterministic) signal in the form of:

$$x(k)=\begin{cases}\sum_{m=1}^{p} B_m z_m^k & \text{for} \quad k\geq 0\\ 0 & \text{for} \quad k<0\end{cases} \qquad [4.26]$$

The parameters of this model belong to two families:

– the "poles" $\{z_m\}$ which carry the frequency f_m and damping α_m information:

$$z_m = e^{-\alpha_m} e^{-j2\pi f_m}$$

– the "complex amplitudes" $\{B_m\}$ which carry the amplitude A_m and phase information ϕ_m:

$$B_m = A_m e^{j\phi_m}$$

We can show that the deterministic signal of equation [4.26] verifies a recursion:

$$x(k) = -\sum_{n=1}^{p} a_n x(k-n) \quad \text{for} \quad k \geq p \qquad [4.27]$$

where the coefficients $\{a_n\}$ are linked to the only poles $\{z_m\}$ in the following manner:

$$\breve{a}(z) \triangleq \sum_{n=0}^{p} a_n z^{-n} = \prod_{m=1}^{p}\left(1 - z_m z^{-1}\right) \qquad [4.28]$$

$$\text{with } a_0 = 1$$

We see that finding the poles $\{z_m\}$ knowing the signal is relatively easy, in two steps:

– from equation [4.27], a group of p values of $x(k)$ help express a linear system whose vector $\mathbf{a} = |a_1 \ldots a_p|^T$ is the solution:

$$\mathbf{a} = -\mathbf{X}^{-1}\mathbf{x}_p \qquad [4.29]$$

$$\mathbf{X} = \begin{pmatrix} x(p-1) & \cdots & x(0) \\ \vdots & \ddots & \vdots \\ x(2p-2) & \cdots & x(p-1) \end{pmatrix} \text{ and } \mathbf{x}_p = \begin{pmatrix} x(p) \\ \vdots \\ x(2p-1) \end{pmatrix}$$

– the expression of the polynomial $\breve{a}(z)$ of equation [4.28] and the solution (numerical) of its roots help obtain the $\{z_m\}$:

$$\mathbf{a} \Rightarrow \breve{a}(z) = 0 \Rightarrow \{z_m\} \qquad [4.30]$$

The resolution of complex amplitudes associated to each pole leads to a simple linear problem (the model of equation [4.26] is linear in $\{B_m\}$ if the $\{z_m\}$ are known):

$$\mathbf{B} = \mathbf{V}^{-1}\mathbf{x}_0 \qquad [4.31]$$

with $\mathbf{V} = \left[z_m^k \right]$ and $\mathbf{x}_0 = |x(o) \ldots x(p-1)|^T$

It can be noted that if we limit some poles to being of unit modulus $|z_m| = 1$, this makes it possible to include the sine waves studied (that is to say, un-dampened) in the model. Such a customization is important for numerous applications, and is closer to the methods presented in Chapter 8. Such a model is known as *harmonic Prony*.

4.3.2. *Noisy deterministic model*

If we now consider a noisy model:

$$\begin{aligned} x(k) &= \sum_{m=1}^{p} B_m z_m^k + n(k) k \geq 0 \\ &= s(k) + n(k) \end{aligned}$$

The previous approach, however, can no longer be applied. But it is legitimate to look for an estimator $\hat{s}(k)$ of $s(k)$. For this, we will use the same steps used previously, but the exact resolutions of linear systems appearing in equations [4.29] and [4.31] will be implemented on the overdetermined systems, in the least-squares sense so as to minimize the influence of the noise term $n(k)$.

From this estimator $s(k)$:

$$\hat{s}(k) \triangleq \sum_{m=1}^{p} \hat{B}_m \hat{z}_m^k$$

will follow a (deterministic) spectral estimator, because:

$$\breve{s}(z) = \sum_{m=1}^{p} \frac{\hat{B}_m}{\left(1 - \hat{z}_m z^{-1}\right)}$$

The relation required from equation [4.2] is thus established.

4.3.3. *Models of random stationary signals*

It is not efficient to apply the approach mentioned above to the implementations of stationary random signals, because the morphology itself of the model is not suitable: a set of decreasing exponentials will be designed for the signals of "transient" form, but unsuitable for the implementation of "permanent" signals. Similarly, the increasing exponentials will not be suitable.

However, the moments or cumulants are deterministic functions, which are made up of decreasing or almost periodic terms: a family of complex exponentials is well suited for this situation.

For example, for a second order analysis, any estimator of the correlation function can be written as:

$$\hat{\gamma}_{xx}(k) = \gamma_{xx}(k) + \varepsilon(k)$$

where $\varepsilon(k)$ is the estimation error (see section 3.2 for its statistical properties), $\gamma_{xx}(\tau)$ has a hermitian symmetry $\gamma_{xx}(-k) = \gamma^*{}_{xx}(k)$ and thus a one-sided model of the following type may be used:

$$\gamma_{xx}(k) = \sum_{m=1}^{p} B_m z_m^k \; k \geq 0$$

completed by symmetry, to gain access to the expression of the PSD (using a bilateral z transform of $\gamma_{xx}(k)$):

$$S_{xx}(z) = \sum_{m=1}^{p} B_m \frac{1 - z_m^2}{\left(1 - z_m z^{-1}\right)\left(1 - z_m z\right)} \qquad [4.32]$$

4.4. Non-linear models

There is a strong interest for this class of models, which help express a very large morphological variability for the signals, much greater than the linear or exponential models. The basic principle remains the same, with a finite dynamic equation:

$$x(k) = F\left[x(k-1), x(k-2)\ldots, x(k-P), u(k),\ldots, u(k-Q), \theta\right]$$

excited by a neutral deterministic or random input.

The models can be characterized by the sub-classes of operators F [.] used, and their relation to the properties of the model. The predictive qualities of these models are mainly the objective of their study.

For the current book, these models present a major difficulty: we generally do not know the relation between the operator F [.], the parameters of the model θ and the spectral characteristics (PSD, etc.). Thus, to date, they are not tools for spectral analysis.

An inquisitive or avid reader can consult the references [PRI 91] and mainly [TON 90].

4.5. Bibliography

[ABR 02] ABRY P., GONÇALVES P., LEVYVÉHEL J., *Lois d'échelle, fractales et ondelettes*, IC2 series, 2 vols., Hermès Sciences, Paris, 2002.

[BEL 89] BELL ANGER M., *Signal Analysis and Adaptive Digital Filtering,* Masson, 1989.

[BOX 70] BOX G., JENKINS G., *Time Series Analysis, Forecasting and Control,* Holden-Day, San Francisco, 1970.

[HLA 05] HLAWATSCH R., AUGER R., OVARLEZ J.-R., *Temps-fréquence,* IC2 series, Hermès Sciences, Paris, 2005.

[KAY 88] KAY S. M., *Modem Spectral Estimation: Theory and Applications,* Prentice Hall, Englewood Cliffs, (NJ), 1988.

[KUM 82] KUMARESAN R., TUFTS D., "Estimating the parameters of exponentially damped sinusoids and pole-zero modeling in noise", *IEEE Trans. Acoust., Speech, Signal Processing,* vol. 30, no. 6, p. 833-840, December 1982.

[LAC 00] LACAZE B., *Processus aléatoires pour les communications numériques,* Hermès Sciences, Paris, 2000.

[MAR 87] MARPLE S., *Digital Spectral Analysis with Applications,* Prentice Hall, Englewood Cliffs (NJ), 1987.

[NAJ 02] NAJIM M., *Traitement numérique des signaux,* Hermès Sciences, Paris, 2002.

[NIK 93] NIKIAS C., PETROPULU A., *Higher-Order Spectra Analysis,* Prentice Hall, 1993.

[PRI 91] PRIESTLEY M., *Non-linear and Non-stationary Time Series Analysis,* Academic Press, 1991.

[TON 90] TONG H., *Non-linear Time Series,* Oxford Science Publications, 1990.

PART II

Non-Parametric Methods

Chapter 5

Non-Parametric Methods

5.1. Introduction

A random discrete time signal $\mathbf{x}(k), k \in Z$ is stationary in the wide sense if its mean m_x and its autocorrelation function $r_{xx}(\kappa)$ defined by:

$$\begin{cases} m_x & = E(\mathbf{x}(k)) \\ r_{xx}(\kappa) & = E\left((\mathbf{x}(k)-m_x)^*(\mathbf{x}(k+\kappa)-m_x)\right) \qquad \forall \kappa \in Z \end{cases} \tag{5.1}$$

are independent of the index k, that is to say independent of the time origin. $\sigma_x^2 = r_{xx}(0)$ is the variance of the considered signal. $\frac{r_{xx}(\kappa)}{\sigma_x^2}$ is the correlation coefficient between the signal at time k and the signal at time $k+\kappa$. Usually we limit ourselves only to the average and the autocorrelation function to characterize a random stationary signal,.even though this second order characterization is very incomplete [NIK 93].

Practically, we have only one realization $x(k), k \in Z$ of a random signal $\mathbf{x}(k)$, for which we can define its time average $\langle x(k) \rangle$:

$$\langle x(k) \rangle = \lim_{N \to \infty} \frac{1}{2N+1} \sum_{k=-N}^{N} x(k) \tag{5.2}$$

Chapter written by Éric LE CARPENTIER.

The random signal $\mathbf{x}(k)$ is known as ergodic for the mean if the mean m_x is equal to the time average of any realization $x(k)$ of this random signal:

$$E\left(x(k)\right)=\left\langle x(k)\right\rangle \text{ ergodicity for the mean} \quad [5.3]$$

Subsequently, we will suppose that the random signal $\mathbf{x}(k)$ is ergodic for the mean and, to simplify, of zero mean.

The random signal $\mathbf{x}(k)$ is ergodic for the autocorrelation if the autocorrelation function $r_{xx}(\kappa)$ is equal to the time average $\left\langle x^*(k)x(k+\kappa)\right\rangle$ calculated from any characterization $x(k)$ of this random signal:

$$E\left(\mathbf{x}^*(k)\mathbf{x}(k+\kappa)\right)=\left\langle x^*(k)x(k+\kappa)\right\rangle \qquad \forall\kappa\in Z \quad [5.4]$$

ergodicity for the autocorrelation

this time average being defined for all values of κ by:

$$\left\langle x^*(k)x(k+\kappa)\right\rangle=\lim_{N\to\infty}\frac{1}{2N+1}\sum_{k=-N}^{N}x^*(k)x(k+\kappa) \quad [5.5]$$

The simplest example of an ergodic stationary random signal for the autocorrelation is the cisoid $ae^{j(2\pi\,\nu_0\,k+\phi)}, k\in Z$ with initial phase ϕ uniformly distributed between 0 and 2π, of the autocorrelation function $a^2e^{j2\pi\,\nu_0\,k}, \kappa\in Z$. However, the ergodicity is lost if the amplitude is also random. In practice, the ergodicity can rarely be rigorously verified. Generally, it is a hypothesis necessary to deduce the second order static characteristics of the random signal considered from a single realization.

Under the ergodic hypothesis, the variance σ_x^2 of the signal considered is equal to the power $\langle|x(k)|^2\rangle$ of all the characterizations x:

$$\sigma_x^2=\left\langle\left|x(k)\right|^2\right\rangle=\lim_{N\to\infty}\frac{1}{2N+1}\sum_{k=-N}^{N}\left|x(k)\right|^2 \quad [5.6]$$

that is to say, the energy of the signal x multiplied by the truncation window $1_{-N,N}$ is equal to 1 over the interval $\{-N, \ldots, N\}$ and zero outside, divided by the length of this interval, when $N\to+\infty$. From Parseval's theorem, we obtain:

$$\begin{aligned}\sigma_x^2&=\lim_{N\to\infty}\frac{1}{2N+1}\int_{-\frac{1}{2}}^{+\frac{1}{2}}\left|\widehat{x1}_{-N,N}(\nu)\right|^2d\nu\\&=\int_{-1/2}^{+1/2}\left\{\lim_{N\to\infty}\frac{1}{2N+1}\left|\widehat{x1}_{-N,N}(\nu)\right|^2\right\}d\nu\end{aligned} \quad [5.7]$$

Thus, from the formula [5.7], we have broken down the power of the signal on the frequency axis using the $\nu \mapsto \lim_{N\to\infty} \frac{1}{2N+1}\left|\widehat{x1}_{-N,N}(\nu)\right|^2$ function. In many books, the spectral density of the power of a random signal is defined by this function. In spite of the ergodic hypothesis, we can show that this function depends on the characterization considered. Here, we will define the power spectral density (or power spectrum) S_{xx} as the overall average of this function:

$$S_{xx}(\nu) = \lim_{N\to\infty} E\left(\frac{1}{2N+1}\left|\widehat{x1}_{-N,N}(\nu)\right|^2\right) \qquad [5.8]$$

$$= \lim_{N\to\infty} E\left(\frac{1}{2N+1}\left|\sum_{k=-N}^{N} x(k)e^{-j\,2\pi\,\nu\,k}\right|^2\right) \qquad [5.9]$$

Thus, we have two characterizations of a stationary random signal in the wide sense, ergodic for the autocorrelation. Wiener-Khintchine's theorem makes it possible to show the equivalence of these two characterizations. Under the hypothesis that the series $\left(\kappa r_{zz}(\kappa)\right)$ is summable, that is:

$$\sum_{\kappa=-\infty}^{+\infty} \left|\kappa r_{zz}(\kappa)\right| < \infty \qquad [5.10]$$

Thus, the power spectral density is the Fourier transform of the autocorrelation function: the two characterizations defined earlier coincide:

$$S_{xx}(\nu) = \hat{r}_{xx}(\nu) \qquad [5.11]$$

$$= \sum_{\kappa=-\infty}^{+\infty} r_{xx}(\kappa)e^{-j\,2\pi\,\nu\,\kappa} \qquad [5.12]$$

This theorem can be demonstrated in the following manner: by expanding the expression [5.9], we obtain:

$$S_{xx}(\nu) = \lim_{N\to\infty} E\left(\frac{1}{2N+1}\sum_{n=-N}^{N}\sum_{k=-N}^{N} \mathbf{x}(n)\mathbf{x}^*(k)e^{-j\,2\pi\,\nu\,(n-k)}\right)$$

From the linearity of the mathematical expectation operator and using the definition of the autocorrelation function:

$$S_{xx}(\nu) = \lim_{N\to\infty} \frac{1}{2N+1} \sum_{n=-N}^{N} \sum_{k=-N}^{N} r_{xx}(n-k) e^{-j\,2\pi\,\nu\,(n-k)}$$

By changing the variable $\kappa = n - k$, the cardinal of the set $\{(n, k) \mid \kappa = n - k$ and $|n| \le N$ and $|k| \le N\}$ is $2N + 1 - |\kappa|$:

$$S_{xx}(\nu) = \lim_{N\to\infty} \frac{1}{2N+1} \sum_{\kappa=-2N}^{2N} \left(2N+1-|\kappa|\right) r_{xx}(\kappa) e^{-j\,2\pi\,\nu\,\kappa}$$

Finally we obtain:

$$\begin{aligned} S_{xx}(\nu) &= \lim_{N\to\infty} \sum_{\kappa=-2N}^{2N} \left(1 - \frac{|\kappa|}{2N+1}\right) r_{xx}(\kappa) e^{-j\,2\pi\,\nu\,\kappa} \\ &= \hat{r}_{xx}(\nu) - \lim_{N\to\infty} \frac{1}{2N+1} \sum_{\kappa=-2N}^{2N} |\kappa|\, r_{xx}(\kappa) e^{-j\,2\pi\,\nu\,\kappa} \end{aligned}$$

Under the hypothesis [5.10], the second term mentioned above disappears, and we obtain the formula [5.12].

These considerations can be taken up succinctly for continuous time signals. A random discrete time signal $\mathbf{x}(t)$, $t \in \Re$ is said to be stationary in the wide sense if its average m_x and its autocorrelation function $r_{xx}(\tau)$ defined by:

$$\begin{cases} m_x = E(\mathbf{x}(t)) \\ r_{xx}(\tau) = E\left((\mathbf{x}(t) - m_x)^* (\mathbf{x}(t+\tau) - m_x)\right) \qquad \forall \tau \in \Re \end{cases} \qquad [5.13]$$

are independent of the time t.

For a characterization $x(t)$, $t \in \Re$ of a random signal $\mathbf{x}(t)$, the time average $\langle x(t) \rangle$ is defined by:

$$\langle x(t) \rangle = \lim_{T\to\infty} \frac{1}{2T} \int_{-T}^{T} x(t)\,dt \qquad [5.14]$$

The ergodicity for the mean can be written as:

$$E\left(x(t)\right)=\left\langle x(t)\right\rangle \qquad [5.15]$$

Subsequently, we will suppose that the random signal $\mathbf{x}(t)$ is ergodic for the mean and, to simplify, of zero mean.

The random signal $\mathbf{x}(t)$ is ergodic for the autocorrelation if:

$$E\left(\mathbf{x}^*(t)\right)\mathbf{x}(t+\tau)=\left\langle x^*(t)x(t+\tau)\right\rangle \qquad \forall\tau\in\Re \qquad [5.16]$$

this time average is defined for all values of τ by:

$$\left\langle x^*(t)x(t+\tau)\right\rangle=\lim_{T\to\infty}\frac{1}{2T}\int_{-T}^{T}x^*(t)x(t+\tau)\,dt \qquad [5.17]$$

The power spectral density S_{xx} can then be expressed by:

$$S_{xx}(f)=\lim_{T\to\infty}E\left(\frac{1}{2T}\left|\widehat{\mathbf{x}\,1_{-T,T}}(f)\right|^2\right) \qquad [5.18]$$

$$=\lim_{T\to\infty}E\left(\frac{1}{2T}\left|\int_{-T}^{T}\mathbf{x}(t)e^{-j\,2\pi f\,t}dt\right|^2\right) \qquad [5.19]$$

If the function $(\tau\, r_{xx}(\tau))$ is summable, that is:

$$\int_{-\infty}^{+\infty}\left|\tau r_{xx}(\tau)\right|d\tau<\infty \qquad [5.20]$$

then the power spectral density is the Fourier transform of the autocorrelation function:

$$S_{xx}(f)=\hat{r}_{xx}(f) \qquad [5.21]$$

$$=\int_{-\infty}^{+\infty}r_{xx}(\tau)e^{-j\,2\pi f\tau}d\tau \qquad [5.22]$$

In both cases (continuous and discrete), we can show that if a linear system invariant over time of pulse response h is excited by a stationary random signal of the autocorrelation function r_{xx} and spectrum S_{xx}, the autocorrelation function r_{yy} and the spectrum S_{yy} of the filter signal $y = h \otimes x$ can be expressed by:

$$r_{yy} = h^{-*} \otimes h \otimes r_{xx} \qquad [5.23]$$

$$S_{yy} = \left|\hat{h}\right|^2 S_{xx} \qquad [5.24]$$

5.2. Estimation of the power spectral density

Under the hypothesis of a continuous or discrete time random signal $\mathbf{x}$ ergodic for the autocorrelation verifying the hypothesis [5.10], and using a recording of finite length $x(t), 0 \leq t \leq T$ or $x(k), 0 \leq k \leq N-1$ of a realization x, we attempt to obtain an estimator of power spectral density S_{xx}, where we can obtain the static characteristics (bias, variance, correlation) as well as the resolution, that is to say the capacity of the estimator to separate the spectral contributions of two narrow band components, for example two cisoids.

The filter bank method (see section 5.2.1), used in analog spectrum analyzers, is more naturally used in the analysis of continuous time signals. The periodogram method (see section 5.2.2) and its variants (see section 5.2.3) are used for the analysis of discrete time signals, and are generally calculated using the fast Fourier transform algorithms (FFT).

5.2.1. *Filter bank method*

The power spectral density describes the distribution of power along the frequency axis. Thus, to estimate this density, it is natural to filter the signal observed $x(t)$ by a band pass filter bank of impulse response h_ℓ whose center frequencies f_ℓ are distributed along the frequency axis. Thus, the power P_ℓ of ℓ^e filter signal $y_\ell = h_\ell \otimes x$ equals:

$$P_\ell = \int_{-\infty}^{+\infty} S_{y_\ell y_\ell}(f)\,df \qquad [5.25]$$

$$= \int_{-\infty}^{+\infty} \left|\hat{h}_\ell(f)\right|^2 S_{xx}(f)\,df \qquad [5.26]$$

If the power spectral density S_{xx} does not vary much in the bandwidth of the ℓ^e filter, then:

$$P_\ell \approx S_{xx}(f_\ell) \int_{-\infty}^{+\infty} \left|\hat{h}_\ell(f)\right|^2 df \qquad [5.27]$$

Thus, the power of the output filter ℓ^e, normalized by the energy of the pulse response of this filter, restores the power spectral density in the center frequency, in the case where this density does not vary much in the bandwidth.

In practice, the frequencies of interest are uniformly distributed along the frequency axis:

$$f_\ell = \ell f_1 \qquad [5.28]$$

and the filters are obtained by spectral offset of the centre low-pass filter on the zero frequency:

$$\hat{h}_\ell(f) = \hat{h}_0(f - f_\ell) \qquad [5.29]$$

This condition being fulfilled, the output power y_ℓ while considering in addition that the transfer function module of the low-pass filter of pulse response h_0 has an even module can be written as:

$$P_\ell = \int_{-\infty}^{+\infty} \left|\hat{h}_0(f_\ell - f)\right|^2 S_{xx}(f)\, df \qquad [5.30]$$

$$= \left(S_{xx} \otimes \left|\hat{h}_0\right|^2\right)(f) \qquad [5.31]$$

To ensure that there is no loss of data, we set a filter bank with perfect reconstruction $\left(x(t) = \sum_\ell y_\ell(t)\right)$, which leads to the following condition:

$$\sum_\ell \hat{h}_\ell(f) = 1 \qquad [5.32]$$

or, something less restrictive, that $x(t)$ and $\sum_\ell y_\ell(t)$ have the same power spectral density, such as the condition:

$$\sum_\ell \left|\hat{h}_\ell(f)\right|^2 = 1 \qquad [5.33]$$

This last condition is more or less fulfilled if the frequential step $f_{\ell+1} - f_\ell = f_1$ is equal to the bandwidth a-3 dB of the filters (interval length $\left\{ f \left| \left| \widehat{h_0}(f) \right| > \frac{\hat{h}(0)}{\sqrt{2}} \right. \right\}$).

From this analysis, we obtain an estimator of power spectral density using a realization $x(t)$, $0 \le t \le T$ from the following sequence of operations:

– *filtering*. The signal $x(t)$, $0 \le t \le T$ is filtered by the bank; we obtain the filter signals $y_\ell(t)$, $0 \le t \le T$;

– *quadration*. We calculate the instantaneous power of these filter signals, that is $|y_\ell(t)|^2$, $0 \le t \le T$;

– *integration*. The power of each filter signal is estimated, without bias, by time averaging the instantaneous power:

$$\overline{P}_\ell = \frac{1}{T} \int_0^T |y_\ell(t)|^2 \, dt \qquad [5.34]$$

– *normalization*. The estimation of the spectral density is obtained by standardizing the estimated power by the energy of the pulse response of the $E_h = \int_{-\infty}^{+\infty} \left| \hat{h}_0(f) \right|^2 df = \int_{-\infty}^{+\infty} h_0^2(t) \, dt$ filters:

$$\overline{S}_{xx}(f_\ell) = \frac{\overline{P}_\ell}{E_h} \qquad [5.35]$$

This estimator is approximately without bias when the spectral density S_{xx} does not vary much in a frequential sampling step. For a sine wave $a\, e^{j(2\pi f_s t + \phi)}$ where ϕ is an initial phase uniformly distributed between 0 and 2π, the spectrum is equal to $a^2 \delta(f - f_s)$; we obtain from formula [5.31] the average value of the spectrum:

$$E\left(\overline{S}_{xx}(f_\ell)\right) = a^2 \frac{\left| \hat{h}_0(f_\ell - f_s) \right|^2}{E_h}$$

The pulse of the exact spectrum leaks to the adjacent analysis frequencies; this phenomenon is more prominent when the frequential sampling step is large. In a signal made up of narrow and large band components, the spectral contribution of the narrow band components can be masked by the spectral contribution of large band components, and all the more when the centre frequency of these narrow band components is farther from an analysis frequency.

5.2.2. *Periodogram method*

Starting from expression [5.9], it is usual to suggest from a recording of N points, the following estimator of the power spectral density known as periodogram:

$$I_{xx}(\nu) = \frac{1}{N}\left|\widehat{x\, 1_{0,N-1}}(\nu)\right|^2 \quad [5.36]$$

$$= \frac{1}{N}\left|\sum_{k=0}^{N-1} x(k)\, e^{-j\,2\pi\,\nu\,k}\right|^2 \quad [5.37]$$

where $1_{0,N-1}$ is the rectangular window equal to 1 over the interval $\{0, \ldots N-1\}$ and zero outside. In relation to the initial definition of the power spectral density, we have lost the mathematical expectation operator, as well as transition to the limit.

From expression [5.11], it is also natural to suggest the following estimator:

$$I'_{xx}(\nu) = \sum_{\kappa=-(N-1)}^{N-1} \bar{r}_{xx}(\kappa) e^{-j\,2\pi\,\nu\,\kappa} \quad [5.38]$$

where $\bar{r}_{xx}$ is a preliminary estimation of the autocorrelation function; the natural estimator of the autocorrelation function is defined by:

$$\bar{r}_{xx}(\kappa) = \begin{cases} \dfrac{1}{N-\kappa}\sum_{k=0}^{N-\kappa-1} x^*(k)\,x(k+\kappa) & \text{if } 0 \le \kappa \le N-1 \\ \dfrac{1}{N+\kappa}\sum_{k=-\kappa}^{N-1} x^*(k)\,x(k+\kappa) & \text{if } -(N-1) \le \kappa \le 0 \\ 0 & \text{otherwise} \end{cases} \quad [5.39]$$

We can easily check that this estimator is non-biased over the interval $|\kappa| \le N-1$. However, we generally prefer using a biased estimator of the autocorrelation function defined by:

$$\bar{r}_{xx}(\kappa) = \begin{cases} \dfrac{1}{N}\sum_{k=0}^{N-\kappa-1} x^*(k)\,x(k+\kappa) & \text{if } 0 \le \kappa \le N-1 \\ \dfrac{1}{N}\sum_{k=-\kappa}^{N-1} x^*(k)\,x(k+\kappa) & \text{if } -(N-1) \le \kappa \le 0 \\ 0 & \text{otherwise} \end{cases} \quad [5.40]$$

The average value of this estimator may be written as:

$$E\left(\bar{r}_{xx}(\kappa)\right) = B_{-N,N}(\kappa)\, r_{xx}(\kappa) \qquad [5.41]$$

where $B_{-N,N}$ is the even Bartlett window, that is:

$$B_{-N,N}(\kappa) = \begin{cases} 1 - \dfrac{|\kappa|}{N} & \text{if } |\kappa| \leq N-1 \\ 0 & \text{otherwise} \end{cases} \qquad [5.42]$$

This estimator, though having a bias, guarantees that the estimation of the Toeplitz hermitian autocorrelation matrix $\left[\bar{r}_{xx}(\ell - k)\right]_{\substack{1\leq\ell\leq N \\ 1\leq k\leq N}}$ is semi-definite positive.

It is easy to show that if the biased estimator of the autocorrelation function is used, then the periodogram and the estimator I'_{xx} are equal.

$$I_{xx} = I'_{xx} \qquad [5.43]$$

From formulae [5.38] and [5.41], we can immediately deduce that the average value of the periodogram is equal to the circular convolution of the exact spectral density and the Fourier transform of the Bartlett window $B_{-N,N}$.

$$E\left(I_{xx}\right) = S_{xx} \otimes \hat{B}_{-N,N} \qquad [5.44]$$

where $B_{-N,N}$ can be expressed by (this result is easily obtained by considering $B_{-N,N}$ as the convolution integral of $(1_{0,N-1}(k))_{k\in Z}$ and $(1_{0,N-1}(-k))_{k\in Z}$ divided by N):

$$\hat{B}_{-N,N}(\nu) = N\left(\frac{\sin(\pi\nu N)}{N\sin(\pi\nu)}\right)^2 \qquad [5.45]$$

This function has a unit integral over a period (by applying Parseval's theorem), and tends towards infinity when N tends towards infinity at integer abscissa. Thus, it tends towards the Dirac comb Ξ_1 of period 1 which is the identity element for the circular convolution integral. Thus, the periodogram is asymptotically unbiased:

$$\lim_{N\to\infty} E\left(I_{xx}\right) = S_{xx} \qquad [5.46]$$

For finite N, the convolution of the real spectrum with Bartlett's window can result in a spectral leakage phenomenon. For example, for a cisoid of amplitude a, the frequency ν_0, of uniformly distributed initial phase between 0 and 2π, the periodogram obtained can be written as $a^2 \hat{B}_{-N,N}(\nu - \nu_0)$. For an addition of two cisoids, the average value of the periodogram obtained is the sum of the contributions of each cisoid. If these two cisoids are of largely different powers, the secondary lobes of Bartlett's window centered on the frequency of the sine wave of high power, that much higher as N is smaller, can mask the main lobe of Bartlett's window centered on the frequency of the sine wave of low power. Thus, the periodogram is less sensitive.

To calculate and draw the periodogram, the frequency axis must be calibrated [0, 1[, for example, using a regular calibration of M points on this interval (with $M \geq N$):

$$\nu \in \left\{ \frac{\ell}{M}, 0 \leq \ell \leq M-1 \right\} \qquad [5.47]$$

Thus, we have, for all $\ell \in \{0, \ldots, M-1\}$:

$$I_{xx}\left(\frac{\ell}{M}\right) = \frac{1}{N} \left| \sum_{k=0}^{N-1} x(k) e^{-j\, 2\pi \frac{\ell k}{M}} \right|^2 \qquad [5.48]$$

By adding $M - N$ zeros to the series of N available points (*zero-padding*), and by choosing for M a power of 2, we calculate this estimator using fast Fourier transform algorithms (FFT: see Chapter 2).

This aspect provides other interpretations of the periodogram. Let us consider the series of M points (that is $(x(0), \ldots, x(N-1))$ completed by $M - N$ zeros) contain a period of a periodic signal x_p of period M. This periodic signal can be broken down into a sum of M cisoids of frequencies $\frac{\ell}{M}$, ℓ varying from 0 to $N-1$, of amplitude $|\hat{x}_p(\ell)|$ and initial phase $\arg(\hat{x}_p(\ell))$:

$$x_p(k) = \sum_{\ell=0}^{M-1} \hat{x}_p(\ell) e^{j\, 2\pi \frac{\ell k}{M}} \qquad [5.49]$$

where $\hat{x}_p$ is the discrete Fourier transform of the series x_p, that is:

$$\hat{x}_p(\ell) = \frac{1}{M} \sum_{k=0}^{N-1} x(k) e^{-j\, 2\pi \frac{\ell k}{M}} \qquad [5.50]$$

$I_{xx}\left(\frac{\ell}{M}\right)$ is thus to the multiplicative factor $\frac{M^2}{N}$ the power of the sine wave ℓ^e present in this decomposition, that is to say, to the factor $\frac{M}{N}$, the average power spectral density of the signal x_p in the interval of a frequency magnitude $\frac{1}{M}$ centered on $\frac{\ell}{M}$:

$$I_{xx}\left(\frac{\ell}{M}\right)=\frac{M}{N}M\left|\hat{x}_p(\ell)\right|^2 \qquad [5.51]$$

It is to be noted that the *zero-padding* technique introduces a redundancy in the periodogram. In fact, knowing only its value for multiple frequencies of $\frac{1}{N}$, we can recreate the estimator of the autocorrelation function from the inverse Fourier transform and, from this estimator, calculate the periodogram at any frequency by the direct Fourier transform. The value of the periodogram at $\frac{\ell}{N}$ frequencies with $0 \leq \ell \leq N-1$ is sufficient. As a result, it is natural to measure the resolution of the periodogram by $\frac{1}{N}$, which is the frequency sampling value of this minimum representation.

The variance of the periodogram is approximately equal to the value of the power spectral density, and does not decrease with the length N of the recording (see section 3.3). The periodogram variants which will be developed in the next paragraph will reduce this variance. As a result, this procedure leads to a deterioration of the resolution power. This resolution/variance compromise, considered as important before obtaining the spectral parametric estimation, is the main disadvantage of standard methods of spectral analysis.

5.2.3. *Periodogram variants*

The average periodogram or Bartlett's periodogram consists of splitting the signal of N points into K sections of length $\frac{N}{K}$, to calculate the periodogram on each segment, and then to average the periodograms obtained:

$$\overline{S}_{xx} = \frac{1}{K}\sum_{k=1}^{K} I_k \quad [5.52]$$

where I_k is the periodogram of the kth segment. If we neglect the dependence between the various segments, we obtain a reduction in the variance by a factor $\frac{1}{K}$ at the cost of a deterioration in the resolution, by $\frac{K}{N}$.

The modified periodogram consists in calculating the periodogram on the signal available $(x(k))_{0 \leq k \leq N-1}$ multiplied by a real truncation window $(w(k))_{k \in Z}$ zero outside the interval $\{0,..., N-1\}$ (see Chapter 2), so as to reduce the spectral leakage and the variance:

$$\overline{S}_{xx} = \frac{1}{NU}\sum_{k=1}^{N} x(k)\,w(k)\,e^{-j\,2\pi\,\nu\,k} \quad [5.53]$$

where U is a standardization factor designed to ensure that the modified periodogram is asymptotically without bias. From the formulae [5.37], [5.38] and [5.43], the following formula can be written:

$$\overline{S}_{xx} = \frac{1}{U}\sum_{\kappa=-(N-1)}^{N-1} \overline{r}_{xx}(\kappa)\,e^{-j\,2\pi\,\nu\,\kappa} \quad [5.54]$$

where $\overline{r}_{xx}$ can be expressed by:

$$\overline{r}_{xx}(\kappa) = \begin{cases} \frac{1}{N}\sum_{k=0}^{N-\kappa-1} x^*(k)\,w(k)\,x(k+\kappa)\,w(k+\kappa) & \text{if } 0 \leq \kappa \leq N-1 \\ \frac{1}{N}\sum_{k=-\kappa}^{N-1} x^*(k)\,w(k)\,x(k+\kappa)\,w(k+\kappa) & \text{if } -(N-1) \leq \kappa \leq 0 \\ 0 & \text{otherwise} \end{cases} \quad [5.55]$$

We immediately obtain by considering $w^-(k) = w(-k)$:

$$E\left(\overline{r}_{xx}(\kappa)\right) = r_{xx}(\kappa)\left(\frac{1}{N} w * w^-\right)(\kappa) \quad [5.56]$$

From the Fourier transform, we obtain:

$$E\left(\overline{S}_{xx}\right) = S_{xx} \otimes \left(\frac{1}{NU}|\hat{w}|^2\right) \qquad [5.57]$$

If the function $\frac{1}{NU}|\hat{w}|^2$ is of unit integral over a period, it tends towards the Dirac comb Ξ_1 and the estimator is asymptotically unbiased. By applying Parseval's theorem, this integral is equal to $\frac{1}{NU}\sum_{k=0}^{N-1} w^2(k)$. Thus, the normalization factor U is equal to:

$$U = \frac{1}{N}\sum_{k=0}^{N-1} w^2(k) \qquad [5.58]$$

This procedure is also accompanied by a loss in resolution, the main lobe of the Fourier transform of all the windows being larger than the rectangular window.

The Welsh periodogram combines the average periodogram and the modified periodogram. We split the signal with N points available into K sections overlapping L points, we calculate the modified periodogram on each segment, then we average the modified periodograms obtained. The deterioration in the resolution is much less and the variance is lower than for $L = 0$; however, the variance does not decrease any further in $\frac{1}{K}$, the different segments are correlated. On a Gaussian sequence, with a Hanning window, we can show that the variance is minimum for $L \approx 0.65K$. Usually $L \approx 0.5K$.

The correlogram, or *windowed periodogram*, consists of a prior estimation of an autocorrelation function using formula [5.38], by applying a real even truncation window w zero outside the interval $\{-(L-1), \ldots, (L-1)\}$ $(L < N)$ and the Fourier transform W:

$$\overline{S}_{xx}(\nu) = \sum_{\kappa=-(L-1)}^{L-1} \overline{r}_{xx}(\kappa)\, w(\kappa)\, e^{-j\,2\pi\,\nu\,\kappa} \qquad [5.59]$$

It is necessary that $w(0) = 1$, so that the power spectral density obtained once again gives the variance $\overline{r}_{xx}(0)$. From the inverse Fourier transform, we obtain:

$$\overline{S}_{xx} = I_{xx} \otimes W \qquad [5.60]$$

The windowed periodogram can be interpreted as a smoothing of the periodogram, which reduces the oscillations caused by the secondary lobes of Bartlett's window and present in the periodogram. If the exact calculation of the variance of the windowed periodogram is difficult, an approximate calculation shows that this variance is approximately multiplied, in relation to the non-windowed periodogram, by the coefficient between 0 and 1 defined by:

$$\frac{1}{N}\sum_{\kappa=-(L-1)}^{L-1} w^2(\kappa)$$

To ensure a positive estimation, it is sufficient to consider a positive Fourier transform truncation window. As a result, Bartlett's window is often used. Blackman and Tuckey, the sources of the method, recommend $L \approx \dfrac{N}{10}$.

As an example, in Figures 5.2 and 5.3 are drawn the power spectral densities estimated using a recording of 128 points of a real value signal sampled at 1 kHz, sum of a sine wave of amplitude 100, frequency 100 Hz, of a sine wave of amplitude 100, frequency 109 Hz, of a sine wave of amplitude 7, frequency 200 Hz, and a white noise of variance 25, and represented in Figure 5.1.

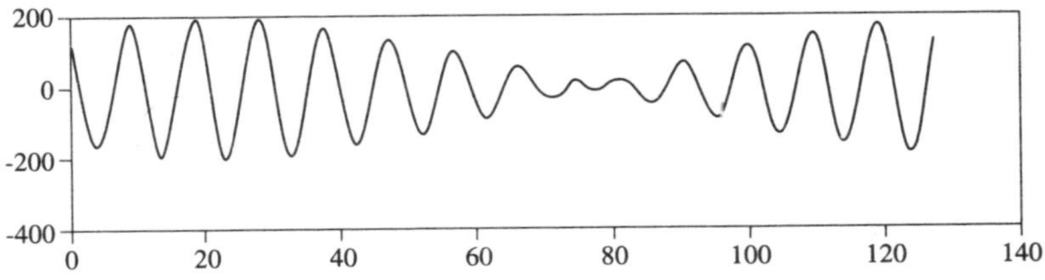

Figure 5.1. *Signal*

In the standard periodogram (Figure 5.2(a)), we mainly observe the lobes of Dirichlet's kernel centered on the frequencies of two predominant sine waves, which mask the spectral contribution of the sine wave of weak amplitude. The use of Hamming's window for the modified periodogram, Figure 5.2(b) reduces the amplitude of the secondary lobes (at the cost of the enlargement of the main lobe, the two predominant sine waves are less apart), and makes it possible to view the contribution of the 3rd sine wave. With a Blackman's window (Figure 5.2(c)), with a larger bandwidth, but with weaker secondary lobes, these two effects are even more prominent. The averaging effect of Bartlett's or Welsh's periodogram on 3 segments of 64 points is overlapped by 32 points, Figure 5.3 reducing the variance, thus the oscillations of the spectrum in the zones describing the bandwidth contributions of white noise, but from the enlargement of the main lobe, the two dominant sine waves cannot be distinguished any more; it is with Bartlett's window, Figure 5.2(c) that we faithfully find the bandwidth contribution of the white noise, close to 14 dB.

5.3. Generalization to higher order spectra

Consider **x** as an ergodic and centered stationary random signal. The moment M_n of order n is defined in the following manner; for all values of $\kappa = [\kappa 1 \ \ldots \ \kappa_{n-1}]^T \in Z_{n-1}$ (also see section 1.2.2.3):

$$M_n(\kappa) = E\left(\mathbf{x}(k)\prod_{i=1}^{n-1}\mathbf{x}(k+\kappa_i)\right) \qquad [5.61]$$

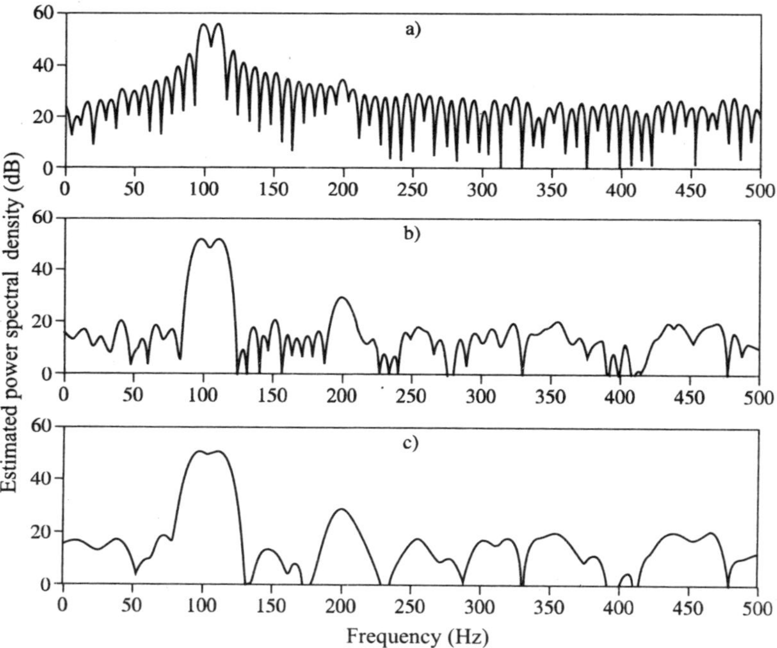

Figure 5.2. *Estimated power spectral density: a) from the periodogram; b) from the modified periodogram (Hamming's window); c) from the modified periodogram (Blackman's window)*

The multispectrum S_n of order n is the multidimensional Fourier transform of the moment of order n: for all values of $\nu = [\nu_1 \cdots \nu_{n-1}]^{\mathrm{T}} \in \Re^{n-1}$:

$$S_n(\nu) = \sum_{\kappa \in Z^n} M_n(\kappa) e^{j2\pi\, \nu^{\mathrm{T}} \kappa} \qquad [5.62]$$

From a recording of finite length $(x(k))_{0 \le k \le N-1}$ of characterization x of a random signal $\mathbf{x}$, the spectrum S_n can be estimated from the natural estimator:

$$\bar{S}_n(\nu) = \sum_{\kappa \in Z^n} \bar{M}_n(\kappa) e^{j\, 2\pi\, \nu^{\mathrm{T}} \kappa} \qquad [5.63]$$

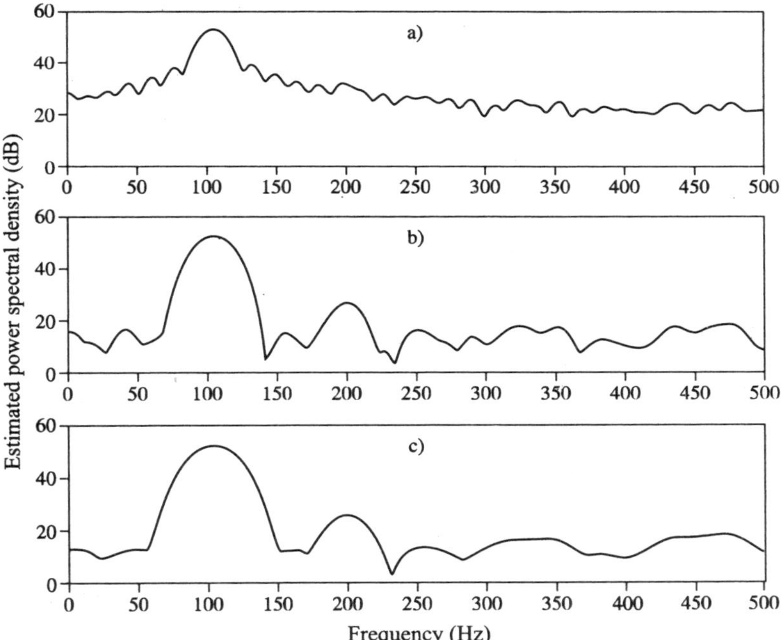

Figure 5.3. *Estimated power spectral density (Hamming's window of 64 samples, covering 32 samples): a) from Bartlett's periodogram; b) from Welsh's periodogram (Hamming's window); c) from Welsh's periodogram (Blackman's window)*

where $\bar{M}_n$ is a biased estimator of the moment of order n:

$$\bar{M}_n(\kappa) = \frac{1}{N} \sum_k x(k) \prod_{i=1}^{n-1} \mathbf{x}(k + \kappa_i) \qquad [5.64]$$

The summation occurring for all values of k between 0 and $N-1$ and such that for all values of I, $k + k_i$ is between 0 and $N-1$. This method (which is indirect because it passes from the estimation of the moment) is similar to the correlogram of the 2nd order. In practice, to reduce the variance of this estimator, we will consider the estimated moment as a multidimensional truncation window w_{n-1}, defined by the product of a standard one-dimensional window w:

$$w_{n-1}(\kappa) = \prod_{i=1}^{n-1} w(\kappa_i) \qquad [5.65]$$

The other method (which is direct because it operates directly on the data), called periodogram of higher order, is defined in the following manner:

$$\overline{S}_n(\nu) = \frac{1}{N} X^*\left(\sum_{i=1}^{n-1} \nu_i\right) \prod_{i=1}^{n-1} X(\nu_i) \qquad [5.66]$$

where $X(\nu_i) = \sum_{k=0}^{N-1} x(k) e^{-j\,2\pi\,\nu\,k}$. As it is of the 2nd order, it is useful to carry out an average of the periodograms from different sections of the original signal, and to weight the different sections by a truncation window – Welsh's procedure (see [NIK 93]).

5.4. Bibliography

[KAY 81] KAY S., MARPLE S., "Spectrum analysis. A modern perspective", *IEEE Proceedings,* vol. 69, no. 11, pp. 1380-1418, 1981.

[MAR 87] MARPLE JR. S., *Digital Spectral Analysis with Applications,* Prentice Hall, 1987.

[MAX 96] MAX J., LACOUME J., *Méthodes et techniques de traitement du signal et applications aux mesures physiques. Volume 1: principes généraux et méthodes classiques,* Masson, 1996.

[NIK 93] NIKIAS C., PETROPULU A., *Higher-order Spectra Analysis,* Prentice Hall, 1993.

[OPP 75] OPPENHEIM A., SCHAEFER R., *Digital Signal Processing,* Prentice Hall, 1975.

[POR 94] PORAT B., *Digital Processing of Random Signals,* Prentice Hall, 1994.

PART III

Parametric Methods

Chapter 6

Spectral Analysis by Stationary Time Series Modeling

Choosing a parametric model among all the existing models is by itself a difficult problem. Generally, this is *a priori* information about the signal and it makes it possible to select a given model. For that, it is necessary to know the various existing models, their properties, the estimation methods for their parameters and the various possible implementations.

In this chapter, we present the various parametric models of stationary processes. Chapter 4 brought an introduction to this chapter by presenting the models of the time series: the ARMA and Prony models have thus already been introduced. The subject of this chapter is to go through these models by tackling their parameter estimation problem as well as their implementation form.

6.1. Parametric models

The stationary parametric models discussed in this chapter can be regrouped into two categories:

– ARMA models,

– Prony models.

Chapter written by Corinne MAILHES and Francis CASTANIÉ.

The "geometrical" methods, based on the decomposition of a signal sub-space and of a noise sub-space (MUSIC, ESPRIT, etc.), specific of sinusoidal signals, are presented in Chapter 8.

These two types of models were presented in Chapter 4, but in order to make the reading of this chapter easier, we will recall the fundamental equations of various models.

As most of the processes can be well approximated by a linear rational model, the ARMA model (p, q) is an interesting general model. It is described by a finite order linear recursion with constant coefficients:

$$x(k) = -\sum_{n=1}^{p} a_n x(k-n) + \sum_{n=0}^{q} b_n u(k-n) \tag{6.1}$$

in which $u(k)$ represents a stationary white noise with null mean.

The coefficients a_n correspond to the AR part of the model, while the coefficients b_n determine the MA part.

A particular case of the ARMA models is the AR model of order p, which is written:

$$x(k) = -\sum_{n=1}^{p} a_n x(k-n) + u(k) \tag{6.2}$$

Autoregressive modeling (AR) is certainly the most popular analysis in parametric spectral analysis. The interest of this method lies in the possibility of obtaining very precise estimations of the AR parameters even for a small number of samples.

In order to have a more physical description of the signal, in terms of frequency, amplitude, phase and damping, the Prony model can also be envisaged. This model consists of describing the signal as a sum of p complex exponentials. This model is thus *a priori* deterministic, contrary to the ARMA and AR models. Thus, it makes it possible to model deterministic signals or moments (autocorrelation functions, for example) by:

$$x(k) = \sum_{m=1}^{p} B_m z_m^k$$

6.2. Estimation of model parameters

In the problem of the estimation of various model parameters, we first take interest in the estimation of AR parameters because this problem is also found again in the ARMA and Prony modelings. We will then tackle the problem of the estimation of the ARMA and Prony model specific parameters.

6.2.1. *Estimation of AR parameters*

Let us suppose that we wish to approach an ordinary signal $x(k)$, for the interval $k = 0, \ldots, N-1$, by the model [6.2]. The method, which is most currently used to calculate the autoregressive coefficients, consists of minimizing the Linear Prediction Error (LPE) in the mean square sense. By considering the recursion equation [6.2], we can realize that we can build a linear predictor of the signal both in the forward and backward direction. We define the forward predictor $\hat{x}(k)$ of $x(k)$ by:

$$\hat{x}(k) = -\sum_{n=1}^{p} a_n x(k-n) \qquad [6.3]$$

We then define the error of forward linear prediction by:

$$e(k) = x(k) - \hat{x}(k) = x(k) + \sum_{n=1}^{p} a_n x(k-n) \qquad [6.4]$$

and the error of backward linear prediction by:

$$b(k) = x(k-p-1) + \sum_{n=1}^{p} a_{p+1-n} x(k-n) \qquad [6.5]$$

The parameters $\mathrm{AR}\{a_n\}_{n=1,p}$ are estimated so that they minimize the following quadratic criterion:

$$\min_{a_k}\left\{\sigma_e^2 = \sum_{k=p}^{N-1} |e(k)|^2\right\} \qquad [6.6]$$

and/or the equivalent backward criterion $\min_{ak}\left\{\sigma_b^2\right\}$.

If the signal is efficiently modeled an AR model of a given order p_1, we can show [KAY 88] that the coefficients calculated this way are good estimators of the coefficients $\{a_n\}$ of equation [6.2] if the order of the chosen model p is equal to p_1.

The solution of this least-squares problem is expressed in the following way:

$$\hat{\mathbf{a}} = -\left(\mathbf{X}^H \mathbf{X}\right)^{-1} \mathbf{X}^H x \text{ with } \hat{\mathbf{a}} = \begin{bmatrix} \hat{a}_1 \\ \vdots \\ \hat{a}_p \end{bmatrix} \quad [6.7]$$

Depending on the chosen minimization window the matrices and the vectors **X** and **x** are defined in different ways. In the case where:

$$\mathbf{X} = \mathbf{X}_1 = \begin{bmatrix} x(p-1) & \cdots & x(0) \\ \vdots & \ddots & \vdots \\ x(N-2) & \cdots & x(N-p-1) \end{bmatrix} \quad \mathbf{x} = \mathbf{x}_1 = \begin{bmatrix} x(p) \\ \vdots \\ x(N-1) \end{bmatrix} \quad [6.8]$$

this estimation method is improperly called the *covariance method* because the matrix $\mathbf{X}_1^H \mathbf{X}_1$ and the vector $\mathbf{X}_1^H \mathbf{x}_1$ are estimations of covariances with one normalizing coefficient. Morf has given an order recursive algorithm making it possible to calculate this solution without explicit inversion of the matrix [KAY 88, MOR 77].

By adopting a least-squares solution, which minimizes the forward [6.6] and backward sum of the prediction errors, we choose:

$$\mathbf{X} = \mathbf{X}_2 = \begin{bmatrix} x^*(N-p) & \cdots & x^*(N-1) \\ \vdots & \ddots & \vdots \\ x^*(p) & \cdots & x^*(1) \\ x(p-1) & \cdots & x(0) \\ \vdots & \ddots & \vdots \\ x(N-2) & \cdots & x(N-p-1) \end{bmatrix} \quad [6.9]$$

$$\mathbf{x} = \mathbf{x}_2 = \begin{bmatrix} x^*(N-p-1) \\ \vdots \\ x^*(0) \\ x(p) \\ \vdots \\ x(N-1) \end{bmatrix}$$

This is the *modified covariance method* and it is generally more efficient than the covariance method. Sometimes it is also called the maximum entropy method

(*MEM*) [LAN 80] because it is a particular case when the noise is Gaussian. Burg developed an order recursive algorithm making it possible to obtain the reflection coefficients that minimize the sum of the forward and backward prediction errors, which makes it possible to deduce the *AR* parameters [BUR 75] *via* the Levinson-Durbin recursion. The advantage of this algorithm is that the estimated poles are always inside or on the unit circle. Its main disadvantages are a spectral line splitting in the case of a signal made up of a noisy sinusoid with a strong signal-to-noise ratio (*SNR*) and sensitivity at the initial phase [SWI 80, WIL 93]. This sensitivity at the phase can be simply highlighted by considering the following case:

$$x(k) = \cos(2\pi fk + \phi) + u(k)$$

$$\hat{\mathbf{a}} = \begin{bmatrix} \hat{a}_1 \\ \hat{a}_2 \end{bmatrix} = -\begin{bmatrix} x(1) & x(0) \\ x(2) & x(1) \end{bmatrix}^{-1} \begin{bmatrix} x(2) \\ x(3) \end{bmatrix} \qquad [6.10]$$

The sensitivity of the solution $\hat{\mathbf{a}}$ is evaluated through the conditioning κ of the system [6.10] [GOL 89]:

$$\kappa = \frac{\lambda_{max}}{\lambda_{min}} \qquad [6.11]$$

or λ_{max} and λ_{min} are the eigenvalues of the following matrix:

$$\begin{bmatrix} \cos(2\pi f + \phi) & \cos(\phi) \\ \cos(4\pi fn + \phi) & \cos(2\pi f + \phi) \end{bmatrix}$$
$$\Rightarrow \begin{cases} \lambda_{max} = \cos(2\pi f + \phi) + \sqrt{\cos(\phi)\cos(4\pi f + \phi)} \\ \lambda_{min} = \cos(2\pi f + \phi) - \sqrt{\cos(\phi)\cos(4\pi f + \phi)} \end{cases} \qquad [6.12]$$

The difference between these two eigenvalues impacts on the extent of the errors in $\hat{\mathbf{a}}$. It is evident that when $\phi = \frac{\pi}{2}$, the eigenvalues are identical and the errors in $\mathbf{a}$ are minimal. When the estimation of the AR parameters is realized in the least-squares sense, the sensitivity at the initial phases of the sinusoids diminishes but it can remain important if the number of samples is small.

But the estimation of the *AR* parameters of an autoregressive model of order *p* can also be done starting from:

$$\gamma_{xx}(m) = E\left[x(k)x^*(k-m)\right] = -\sum_{n=1}^{p} a_x \gamma_{xx}(m-n) + \sigma_n^2 \delta(m) \qquad [6.13]$$

which we call *Yule-Walker* equations, which were already mentioned in Chapter 4. This result shows that the autocorrelation of $x(k)$ satisfies the same recursion as the signal. A large number of estimation methods solve the Yule-Walker equations by replacing the theoretic autocorrelation $\gamma_{xx}(m)$ by an estimation $\hat{\gamma}_{xx}(m)$:

$$\hat{\mathbf{a}} = -\underbrace{\begin{bmatrix} \hat{\gamma}_{xx}(0) & \hat{\gamma}_{xx}^*(1) & \cdots & \hat{\gamma}_{xx}^*(p-1) \\ \hat{\gamma}_{xx}(1) & \hat{\gamma}_{xx}(0) & \cdots & \hat{\gamma}_{xx}^*(p) \\ \vdots & \ddots & \ddots & \vdots \\ \hat{\gamma}_{xx}(p-1) & \cdots & \cdots & \hat{\gamma}_{xx}(0) \end{bmatrix}^{-1}}_{\hat{\mathbf{R}}_x^{-1}} \underbrace{\begin{bmatrix} \hat{\gamma}_{xx}(1) \\ \hat{\gamma}_{xx}(2) \\ \vdots \\ \hat{\gamma}_{xx}(p) \end{bmatrix}}_{\hat{\mathbf{r}}_x} \qquad [6.14]$$

It is the *autocorrelation method* and $\hat{\mathbf{R}}_x$ is the (estimated) autocorrelation matrix. When $\hat{\gamma}_{xx}(m)$ is the biased estimator of the correlation, the poles are always inside the unit circle, which is not the case with the non-biased estimator which gives, however, a better estimation. The Levinson-Durbin algorithm [LEV 47] provides an order recursive solution of the system [6.14] in $O(p^2)$ computational burden. In order to reduce the noise influence and to obtain a better estimation, the system [6.14] can use a larger number of equations (>> p) and be solved in the least-squares sense (LS) or in the total least-squares sense (TLS) [VAN 91]. These methods are called LSYW and TLSYW (see [DUC 98]):

$$\text{LSYW} \quad \hat{\mathbf{a}} = -\left(\hat{\mathbf{R}}_x^H \hat{\mathbf{R}}_x\right)^{-1} \hat{\mathbf{R}}_x^H \hat{\mathbf{r}}_x$$

$$\text{TLSYW} \quad \hat{\mathbf{a}} = -\left(\hat{\mathbf{R}}_x^H \hat{\mathbf{R}}_x - \sigma_{\min}^2 I\right)^{-1} \hat{\mathbf{R}}_x^H \hat{\mathbf{r}}_x \qquad [6.15]$$

$$\hat{\mathbf{R}}_x \begin{bmatrix} \hat{\gamma}_{xx}(0) & \cdots & \hat{\gamma}_{xx}^*(p-1) \\ \vdots & \ddots & \vdots \\ \hat{\gamma}_{xx}(N_0-1) & \cdots & \hat{\gamma}_{xx}(N_0-p) \end{bmatrix} \quad \hat{\mathbf{r}}_x = \begin{bmatrix} \hat{\gamma}_{xx}(1) \\ \vdots \\ \hat{\gamma}_{xx}(N_0) \end{bmatrix}$$

where $\sigma_{\min}$ is the smaller singular value of the matrix $\left[\hat{\mathbf{R}}_x \, \hat{\mathbf{r}}_x\right]$, $\boldsymbol{I}$ the identity matrix and N_0 the number of equations. The value of N_0 should be at most of the order of $\frac{N}{2}$ when the non-biased estimator of the correlation is used so as not to make correlations with a strong variance intervene.

The TLS solution is more efficient than the LS method because it minimizes the errors in $\hat{\mathbf{r}}_x$ and in $\hat{\mathbf{R}}_x$ at the same time, but it presents a disadvantage which we will encounter later. It is generally better to envisage a calculation of the LS and TLS solutions via the singular value decomposition (SVD) of the matrix $\hat{\mathbf{R}}_x$, particularly when the system is not well conditioned, because the SVD proves a greater numeric stability. The LS and TLS solutions calculated by the SVD are:

$$\hat{\mathbf{R}}_x = \hat{\mathbf{U}}_p \hat{\boldsymbol{\Sigma}}_p \hat{\mathbf{V}}_p^H \quad \hat{\boldsymbol{\Sigma}}_p = diag\left(\hat{\sigma}_1, \cdots, \hat{\sigma}_p\right) \quad \hat{\sigma}_k \geq \hat{\sigma}_{k+1}$$

$$\begin{aligned} &\text{LSYW} \quad \hat{\mathbf{a}} = -\hat{\mathbf{V}}_p \left(\hat{\boldsymbol{\Sigma}}_p\right)^{-1} \hat{\mathbf{U}}_p^H \hat{\mathbf{r}}_x \quad \left(\hat{\boldsymbol{\Sigma}}_p\right)^{-1} = diag\left(\tfrac{1}{\hat{\sigma}_1}, \ldots, \tfrac{1}{\hat{\sigma}_p}\right) \\ &\text{TLSYW} \quad \begin{bmatrix} \hat{\mathbf{a}} \\ 1 \end{bmatrix} = \frac{\mathbf{v}_{p+1}}{\mathbf{v}_{p+1}(p+1)} \end{aligned} \qquad [6.16]$$

where $\mathbf{V}_{p+1}$ is the eigenvector associated to the smallest eigenvalue of the matrix $\left[\hat{\mathbf{R}}_x \, \hat{\mathbf{r}}_x\right]^H \left[\hat{\mathbf{R}}_x \, \hat{\mathbf{r}}_x\right]$, i.e. the $(p + 1)^{\text{th}}$ column vector of the matrix $\hat{\mathbf{V}}_{p+1}$ so that $\left[\hat{\mathbf{R}}_x \, \hat{\mathbf{r}}_x\right] = \hat{\mathbf{U}}_{p+1} \hat{\boldsymbol{\Sigma}}_{p+1} \hat{\mathbf{V}}_{p+1}^H$.

There exists an extension of the Levinson-Durbin algorithm for solving LSYW, it is what we call the least-squares *lattice* [MAK 77, PRO 92]. The interest of this algorithm is that it is time and order recursive, which makes it possible to implement parameter estimation adaptive procedures.

We can very easily notice that these methods are biased because the correlation at 0th lag makes the power of the white noise σ_u^2 intervene. The approached value of this bias for (6.14) is:

$$\hat{\mathbf{a}} = -\left(\hat{\mathbf{R}}_{x-u} + \sigma_u^2 I\right)^{-1} \hat{\mathbf{r}}_{x-u}$$

$$\hat{\mathbf{a}} \approx \mathbf{a} + \sigma_u^2 \hat{\mathbf{R}}_{x-u}^{-2} \hat{\mathbf{r}}_{x-u} = \mathbf{a} - \sigma_u^2 \hat{\mathbf{R}}_{x-u}^{-1} \mathbf{a} \qquad [6.17]$$

where $x - u$ is the noiseless signal. Certain methods [KAY 80, SAK 79] exploit this relation in order to try to eliminate this bias, but these require an estimation of σ_u^2 The simplest solution for obtaining a non-biased estimation of the *AR* parameters consists of not making the zero lag correlation intervene in the estimation:

$$\hat{\mathbf{a}} = -\begin{bmatrix} \hat{\gamma}_{xx}(p) & \cdots & \hat{\gamma}_{xx}(1) \\ \vdots & \ddots & \vdots \\ \hat{\gamma}_{xx}(2p-1) & \cdots & \hat{\gamma}_{xx}(p) \end{bmatrix}^{-1} \begin{bmatrix} \hat{\gamma}_{xx}(p+1) \\ \vdots \\ \hat{\gamma}_{xx}(2p) \end{bmatrix} \qquad [6.18]$$

That is called the modified Yule-Walker method (MYW). When we solve this system in the classic or total least-squares sense, we call these methods LSMYW and TLSMYW [STO 92]. When the signal $x(k)$ is really an AR(p), these estimators are asymptotically unbiased ($N \to \infty$). However, if $x(k)$ is made up of a sum of sinusoids and white noise, the MYW estimators are also asymptotically unbiased [GIN 85].

The case of a colored noise slightly modifies the MYW estimators. The hypothesis that the noise $u(k)$ is of the form of a moving average process (MA) of order q is the most common:

$$u(k) = \sum_{n=0}^{q} b_n \varepsilon(k-n), \qquad [6.19]$$

where $\varepsilon(n)$ is a null mean Gaussian white noise. In this case, the correlation $\gamma_{xx}(m)$ of $u(k)$ is null when $m > q$. Thus, an estimator of the *AR* parameters of an ARMA (p, q) or of a sum of noisy sinusoids by a MA(q) is:

$$\hat{\mathbf{a}} = -\begin{bmatrix} \hat{\gamma}_{xx}(p+q) & \cdots & \hat{\gamma}_{xx}(q+1) \\ \vdots & \ddots & \vdots \\ \hat{\gamma}_{xx}(2p+q-1) & \cdots & \hat{\gamma}_{xx}(p+q) \end{bmatrix}^{-1} \begin{bmatrix} \hat{\gamma}_{xx}(p+q+1) \\ \vdots \\ \hat{\gamma}_{xx}(2p+q) \end{bmatrix} \qquad [6.20]$$

Of course, identifying the noise structure in practice is far from being evident and very often we satisfy ourselves with the whiteness hypothesis.

These methods suppose that the order p is known. In practice it is not necessarily simple to determine it (we will see the methods further) but it should not be underestimated in order not to forget to estimate all the signal poles. When it is overestimated, besides the signal poles, poles linked to the noise also appear; they are more or less damped but in any case they are generally more damped than the signal poles (if the SNR is not too weak). This makes it possible to distinguish between the signal poles and the noise poles. The LSMYW method of an overestimated order is called high-order Yule-Walker method (HOYW). An order overestimation presents the advantage of improving the signal poles estimation. The LSMYW method is relatively efficient in order to distinguish between the signal and the noise poles, on the contrary the TLSMYW method has the disadvantage of attracting the poles (including those of the noise) on the unit circle making the distinction very difficult. We will show this phenomenon on one example; we consider N = 100 samples of a signal made up of non-damped sinusoids of frequencies 0.2 and 0.3 and of a white noise such as SNR = 10 dB. The poles are estimated by the LSMYW and TLSMYW methods with a number of equations $\frac{N}{2}$

and an order p = 10. Figure 6.1 represents the plot of the poles estimated in the complex plane for a signal realization.

We notice that the poles linked to the noise of the LSMYW method are clearly damped in relation to those obtained with TLSMYW which sometimes fall outside the unit circle. When the order is overestimated, the TLSMYW method should be avoided. However, for a correct order, the estimation by TLSMYW is better than that of LSMYW.

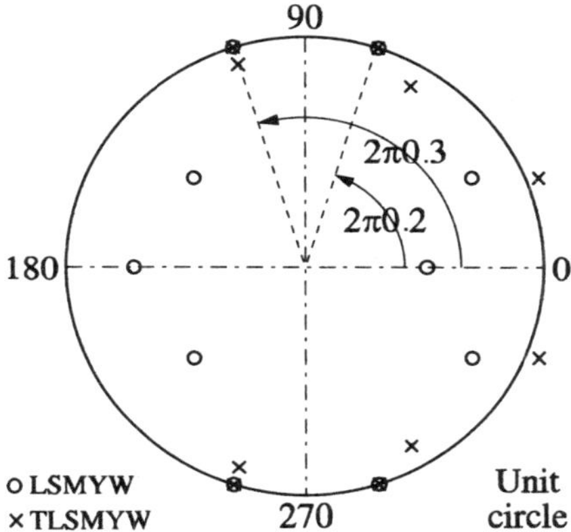

Figure 6.1. *Influence of the order overestimation for the LSMYW and TLSMYW methods*

Up to now, the LSMYW method is one of the most efficient of those that we have described [CHA 82, STO 89]. Stoica and Söderström [SÖD 93] have given the asymptotic performances of this method in the case of a noisy exponential (white noise) of circular frequency $\omega = 2\pi f$ and amplitude A:

$$Var(\hat{\omega}) = \lim_{N\to\infty} E\left[(\hat{\omega}-\omega)^2\right] = \frac{1}{N}\frac{\sigma_u^4}{A^4}\frac{2(2p+1)}{3p(p+1)N_0^2} \quad [6.21]$$

where p represents the order and $N_0 \geq p$ the number of equations. This expression can be compared to the asymptotic Cramer-Rao bound of the circular frequency ω (see section 3.4.1):

$$CRB(\omega) = \frac{6\sigma_u^2}{A^2 N^3} \quad [6.22]$$

Or we could think that the variance of HOYW [6.21] can become smaller than the Cramer-Rao bound. Actually, this is not the case at all if it is only asymptotically that it can become very close to Cramer-Rao. [SÖD 91] compares this method to the Root-MUSIC and ESPRIT methods (see Chapter 8) in terms of precision and computational burden. HOYW makes a good compromise between these two aspects.

6.2.2. *Estimation of ARMA parameters*

The estimation of ARMA parameters [KAY 88, MAR 87] is made in two steps: first, we estimate the AR parameters and then the MA parameters. This suboptimal solution leads to a considerable reduction of the computational complexity with respect to the optimal solution that would consist of estimating the AR and MA parameters. The estimation can be efficiently performed in two steps. Starting from equation [6.1], we show that the autocorrelation function of an ARMA process itself follows a recursion of the AR type but starting from the lag $q+1$ (see section 4.2.2, equation [4.17]):

$$\gamma_{xx}(m) = -\sum_{n=1}^{p} a_x \gamma_{xx}(m-n) \text{ for } m > q$$

The estimation of the AR parameters is done as before by taking these modified Yule-Walker equations into account (starting from the rank $q + 1$).

The estimation of the MA parameters is done first by filtering the process by the inverse AR filter, using the AR estimated parameters so that we come back to an MA of order q, according to the principle in Figure 6.2.

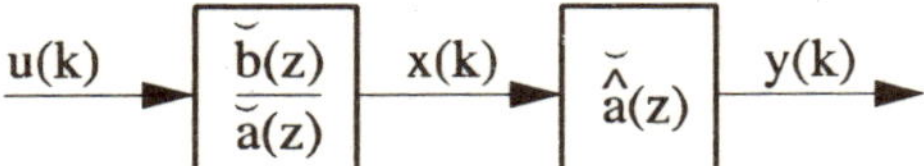

Figure 6.2. *Principle of the estimation of MA parameters in an ARMA*

If the inverse AR filter is supposed perfectly known, we obtain (by using equation [6.1]):

$$y(k) = x(k) + \sum_{n=1}^{p} a_n x(k-n) = \sum_{n=0}^{q} b_n u(k-n)$$

and the filtered process $y(k)$ is a pure MA(q). Thus, we purely come back to a problem of estimation of MA parameters. This model MA(q) can theoretically be modeled by an infinite order AR by writing:

$$\breve{b}(z)=\frac{1}{\breve{c}(z)} \text{ with } \breve{c}(z)=\sum_{n=0}^{+\infty} c_n z^{-n}$$

$$c_k = -\sum_{n=1}^{q} b_n c_{k-n} + \delta(k)$$

We obtain a relation between the MA coefficients and the coefficients of the equivalent AR model:

$$c_k = -\sum_{n=1}^{q} b_n c_{k-n} + \delta(k)$$

$\delta(k)$ standing for the Kronecker symbol ($\delta(0) = 1$ and $\delta(k) = 0$, for any $k \neq 0$).

In practice, we choose to estimate an AR model of high order M such as $M >> q$. By using the AR parameters estimated this way, we obtain a set of equations in the form:

$$\varepsilon(k) = \hat{c}_k + \sum_{n=1}^{q} b_n \hat{c}_{k-n} \qquad [6.23]$$

Ideally, $\varepsilon(k)$ should be null everywhere except for $k = 0$ where it should be equal to 1. As the order of the estimated AR is not infinite, it's nothing of the sort and the estimation of the MA parameters should be done by minimizing a quadratic criterion of the form:

$$\sum_k |\varepsilon(k)|^2$$

The index k varies on a domain, which differs according to the used estimation methods. Indeed, equation [6.23] is not unlike the linear prediction error of an AR model in which the parameters would be the b_n and the signal $\hat{c}_k$. From this fact, the AR estimation techniques can be envisaged with, as particular cases: $k = 0, ..., M + q$ corresponding to the method of autocorrelations and $k = q, ..., M$ corresponding to the covariance method.

6.2.3. *Estimation of Prony parameters*

The classic estimation method of the Prony model parameters is based on the recursion expression of a sum of exponentials:

$$x(k) = \sum_{n=1}^{p} B_n z_n^k = -\sum_{n=1}^{p} a_n x(k-1) \text{ for } k \geq p \qquad [6.24]$$

which leads to:

$$x(k) = \sum_{n=1}^{p} a_n x(k-n) + \sum_{n=0}^{p} a_n u(k-n) \quad p \leq k \leq N-1 \quad (a_0 = 1) \qquad [6.25]$$

We find the well-known equivalence between a sum of p noisy exponentials and the ARMA(p, p) model for which the AR and MA coefficients are identical. The poles z_n are deduced from the polynomial roots:

$$\breve{a}(z) = \sum_{n=0}^{p} a_n z^{-n} = \prod_{n=1}^{p} \left(1 - z^{-1} z_n\right) \qquad [6.26]$$

The classic estimation method of the AR parameters in the estimation procedure of the Prony model parameters consists of minimizing the quadratic error:

$$e = \min_{a_n} \sum_{k=p}^{N-1} \left| x(k) + \sum_{n=1}^{p} a_n x(k-n) \right|^2 \qquad [6.27]$$

which, in a matrix form, is written as:

$$\min_{a_n} \left\| \begin{bmatrix} x(p-1) & \cdots & x(0) \\ \vdots & \ddots & \vdots \\ x(N-1) & \cdots & x(N-p) \end{bmatrix} \begin{bmatrix} a_1 \\ \vdots \\ a_p \end{bmatrix} + \begin{bmatrix} x(p) \\ \vdots \\ x(N) \end{bmatrix} \right\|^2$$

$$\min_{\mathbf{a}} \left\| \mathbf{X}\mathbf{a} + \mathbf{x} \right\|^2 \qquad [6.28]$$

and leads to the least-squares solution $\mathbf{a}_{LS}$:

$$\mathbf{a}_{LS} = -\left(\mathbf{X}^H \mathbf{X}\right)^{-1} \mathbf{X}^H \mathbf{x}. \qquad [6.29]$$

This method is sometimes called *LS-Prony*. When the system is solved in the total least-squares sense, we speak of *TLS-Prony*. We will note that the matrix $\mathbf{X}^H\mathbf{X}$ is the estimation of the covariance matrix, with one multiplying factor ($\frac{1}{N-p}$). This method is thus identical to the covariance method for the estimation of the *AR* coefficients and of the signal poles. This method can be slightly changed in the case of $\frac{p}{2}$ non-damped real sinusoids to force the estimated poles to be of unit module:

$$\min_{a_n} \left\| \begin{bmatrix} x^*(N-p+1) & \cdots & x^*(N) \\ \vdots & \ddots & \vdots \\ x^*(1) & \cdots & x^*(p) \\ x(p-1) & \cdots & x(0) \\ \vdots & \ddots & \vdots \\ x(N-1) & \cdots & x(N-p) \end{bmatrix} \begin{bmatrix} a_1 \\ \vdots \\ a_{\frac{p}{2}-1} \\ a_{\frac{p}{2}} \\ a_{\frac{p}{2}-1} \\ \vdots \\ a_1 \\ 1 \end{bmatrix} + \begin{bmatrix} x^*(N-p) \\ \vdots \\ x^*(0) \\ x(p) \\ \vdots \\ x(N) \end{bmatrix} \right\|^2 \qquad [6.30]$$

We speak then of the *harmonic Prony* method. Other estimation methods of the *AR* parameters, implementing the correlation, can be used: *LSYW* and *LSMYW*, for example.

Once the poles are estimated $(\hat{z}_n)$, the complex amplitudes are obtained by solving the Vandermonde system:

$$\begin{bmatrix} 1 & 1 & \cdots & 1 \\ \hat{z}_1 & \hat{z}_2 & \cdots & \hat{z}_p \\ \hat{z}_1^2 & \hat{z}_2^2 & \cdots & \hat{z}_p^2 \\ \vdots & \vdots & \ddots & \vdots \\ \hat{z}_1^{M-1} & \hat{z}_2^{M-1} & \cdots & \hat{z}_p^{M-1} \end{bmatrix} \begin{bmatrix} B_1 \\ \vdots \\ B_p \end{bmatrix} \approx \begin{bmatrix} x(0) \\ x(1) \\ x(2) \\ \vdots \\ x(M-1) \end{bmatrix} \qquad [6.31]$$

$$\hat{\mathbf{V}}\mathbf{B} \approx \mathbf{x}$$

Solutions in the least-squares sense and in the total least-squares sense can be envisaged. It was shown in [DUC 97] that the total least-squares solution is less efficient in terms of bias than that of the classic least-squares. On the other hand, the number of equations M of the Vandermonde system can be chosen in an optimal manner, as detailed in [DUC 95].

6.2.4. *Order selection criteria*

Choosing a model system is a problem which is as important as the choice of the model itself. Selecting too small an order means smoothing the obtained spectrum, while choosing too large an order introduces secondary spurious peaks.

There is a large number of order selection criteria which are for most cases based on the statistic properties of the signal (maximum likelihood estimation: MLE). Others, simpler and less efficient are based on the comparison of the eigenvalues of the correlation matrix to some threshold correlation matrix [KON 88].

A large number of order selection criteria use the prediction error power decrease when the order increases. When the theoretic order is reached, this power remains constant. However, a criterion based only on the prediction error power shape does not make it possible to take the estimated spectrum variance increase into account when the order is overestimated. That is why the criteria integrate these two phenomena. One of the first criteria proposed by Akaike [AKA 70] was the *FPE* (Final Prediction Error): the estimated error corresponds to the value that minimizes:

$$FPE(k) = \frac{N+k}{N-k}\hat{\rho}_k \qquad [6.32]$$

where:

$$\hat{\rho}_k = \hat{\gamma}_{xx}(0) + \sum_{l=1}^{k} \hat{a}_l \hat{\gamma}_{xx}(l) \qquad [6.33]$$

is the power of the prediction error that decreases with k while the term $\frac{N+k}{N-k}$ increases with k (to take the estimated spectrum variance augmentation into account when k increases). The AR parameters are estimated through Yule-Walker equations with the biased estimator of the correlation. The most well known criterion proposed by Akaike is the AIC (Akaike Information Criterion) [AKA 74]:

$$AIC(k) = N\ \ln(\hat{\rho}_k) + 2k \qquad [6.34]$$

This criterion is more general than FPE and it can be applied by determining the order of an MA part of an ARMA model. Asymptotically $(N \to \infty)$ FPE and AIC are equivalent, but for a small number of samples AIC is better. It was proved that AIC is inconsistent and it tends to overestimate the order [KAS 80]. [RIS 83] proposed to modify AIC by replacing the term *2k* by a term, which increases more rapidly (depending on N) k ln (N). This criterion is named MDL (Minimum Description Length):

$$\mathrm{MDL}(k) = N\ \ln(\hat{\rho}_k) + k\ln(N)\ . \qquad [6.35]$$

This criterion is consistent and gives better results than AIC [WAX 85].

It would be tedious to present all the criteria that were developed; for more information on this, see the following references: [BRO 85, BUR 85, FUC 88, PUK 88, YIN 87, WAX 88].

[WAX 85] expressed the AIC and MDL criteria depending on the eigenvalues of the autocorrelation matrix $\hat{R}_y$:

$$AIC(k) = N(p-k)\ln\left(\frac{\left(\prod_{t=k+1}^{p} \hat{\lambda}_t\right)^{\frac{1}{p-k}}}{\frac{1}{p-k}\sum_{t=k+1}^{p} \hat{\lambda}_t}\right) + k(2p-k) \quad k = 1,\cdots,p-1 \qquad [6.36]$$

$$MDL(k) = N(p-k)\ln\left(\frac{\left(\prod_{t=k+1}^{p} \hat{\lambda}_t\right)^{\frac{1}{p-k}}}{\frac{1}{p--k}\sum_{t=k+1}^{p} \hat{\lambda}_t}\right) + \frac{1}{2}k(2p-k)\ln(N) \qquad [6.37]$$

where $\hat{\lambda}_t$ are the ordered eigenvalues $(\hat{\lambda}_t \geq \hat{\lambda}_{t+1})$ of the matrix $\hat{R}_v(p \times p)$. It is possible to define these criteria according to the singular values $\hat{\sigma}_t$ of the matrix $\hat{R}_v(M \times p, M > p)$ by replacing $\hat{\lambda}_t$ by $\hat{\sigma}_t^2$ in the matrices [6.36] and [6.37] [HAY 89].

The increase in the number of lines of the matrix $\hat{R}_v$ evidently makes an improvement of the order estimation performances possible. We have compared the two expressions [6.35] and [6.37] of the MDL criterion on the following example. We consider N = 100 samples of a signal made up of two sinusoids of identical amplitudes of frequencies 0.1 and 0.2 and of one white noise. The dimension of the matrix $\hat{R}_v$ is (30 x 30). The simulations were carried out for signal-to-noise ratios of 10 dB and 0 dB and are presented in Figures 6.3 and 6.4.

SNR = 10 dB (Figure 6.3): the dimension of the subspace signal is 4 because the signal has two sinusoids. The criterion [6.37] (top curve) reaches its minimum for $k = 4$. However, the criterion [6.35] (bottom curve) is minimum when $k = 8$. These results illustrate the efficiency of the criterion defined starting from the eigenvalues of the autocorrelation matrix and the mediocre performances of the criterion defined starting from the prediction error power.

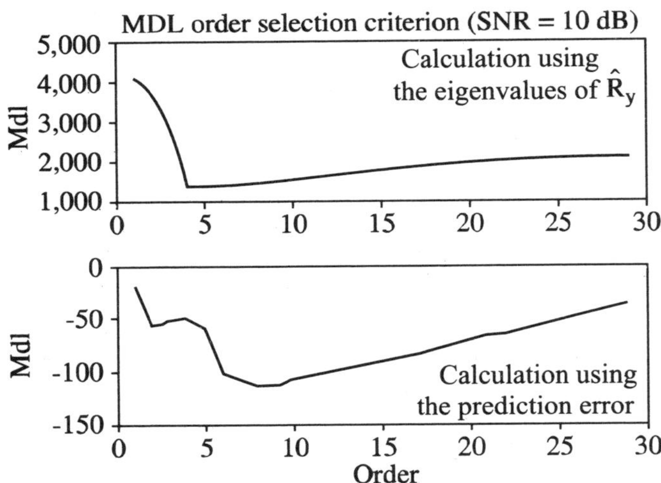

Figure 6.3. *Comparison of the MDL criteria [6.35] and [6.37] for SNR = 10 dB*

SNR = 0 dB (Figure 6.4): for a weaker SNR, the criterion [6.35] (top curve) is inefficient while the criterion [6.37] (top curve) gives a correct result.

These simulations have highlighted the efficiency of the MDL criterion [6.37] and, in a more general manner, the efficiency of the criteria built starting from the eigenvalues of the autocorrelation matrix $\hat{\mathbf{R}}_x$.

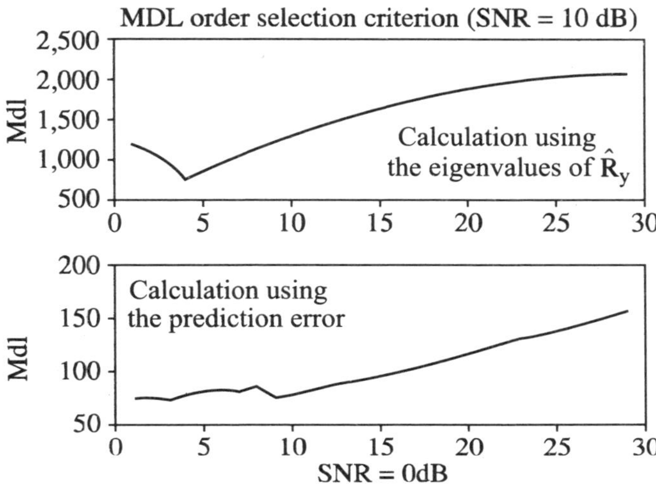

Figure 6.4. *Comparison of the MDL criteria [6.35] and [6.37] for SNR = 0 dB*

In the case of a Prony model, the model order corresponds to the number of searched exponential components, which evidently reduces to an order determination problem of AR model. All the preceding criteria are applicable.

In the case of an ARMA model, selecting the models p and q of the respective AR and MA parts is not a simple problem. Few published works tackle this problem, excepting the very simple cases. The AIC criterion is one of the most used in the form:

$$\text{AIC}(\text{p,q}) = N \ln\left(\hat{\rho}_{pq}\right) + 2(p+q)$$

where $\hat{\rho}_{pq}$ represents the estimated power of the entrance noise of the ARMA model. As in the case of the AR model, the minimum of this function with two variables provides the couple (p, q) of the AR and MA orders to be taken into account.

6.3. Properties of spectral estimators produced

In the case of a signal made up of a sinusoid and an additive white noise, we take interest in the AR model spectrum evolution, according to the signal to noise ratio (SNR), for a fixed AR order. Various results can be found in the literature, depending on the hypotheses taken for the modeled signal. For an example, we can mention [LAC 84] which provides the equivalent bandwidth expression for a pure sinusoid with additive white noise, modeled by an AR model of order p:

$$\Delta f_p = \frac{6}{\pi p(p+1)\beta}$$

β being the signal to noise ratio. Figure 6.5 makes it possible to highlight that the bandwidth at -3dB of the spectrum lobe depends on the SNR: the weaker the SNR is, the more the lobe enlarges. The AR spectral resolution is affected by the noise presence.

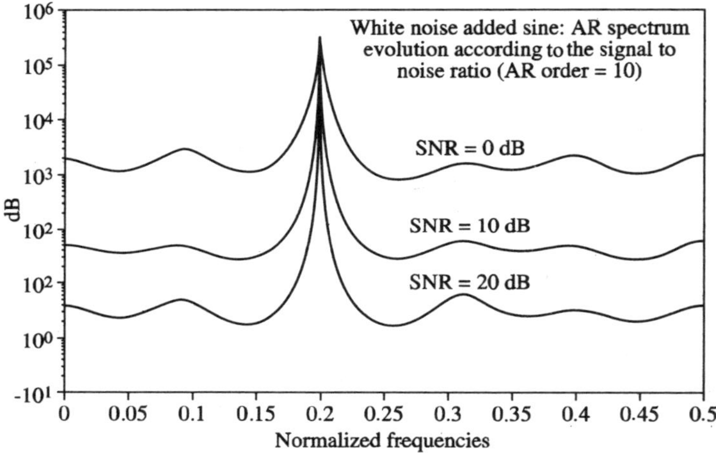

Figure 6.5. *Evolution of the AR spectrum according to the SNR in the case of a noisy sinusoid*

In the case of a noisy sinusoidal signal, Figure 6.6 plots the evolution of the AR spectrum according to the order of the chosen model, for a previously fixed SNR. The larger the order is, the better the spectral resolution is (inversely related to the width of the "peaks"). But if the chosen order is too big, false (or spurious) "peaks" appear.

The effect of the order choice is again illustrated on the case of a signal made up of three sinusoids in white noise. The minimal order necessary for the estimation of these three sinusoids is of 6. Of course, the presence of the additive white noise makes it necessary to choose a higher order. Figure 6.7 presents the AR spectrum evolution, while the chosen order is below 6: if we could see that like an animated cartoon, we would see the successive lines "push" according to the chosen order. Below, we have represented the periodogram, which makes it possible to appreciate the smoothing produced by the AR spectral analysis.

Figure 6.8 presents the AR spectrum evolution on the same signal made up of three sinusoids in white noise when the chosen order is superior to 6. The three lines are better and better estimated but at the order 50, the AR modeling makes two false "peaks" appear, which are actually noise peaks.

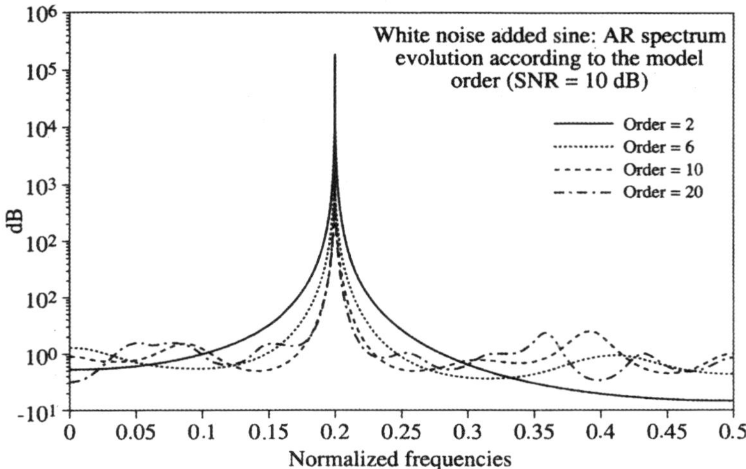

Figure 6.6. *AR spectrum evolution according to the order of the chosen model in the case of a noisy sinusoid*

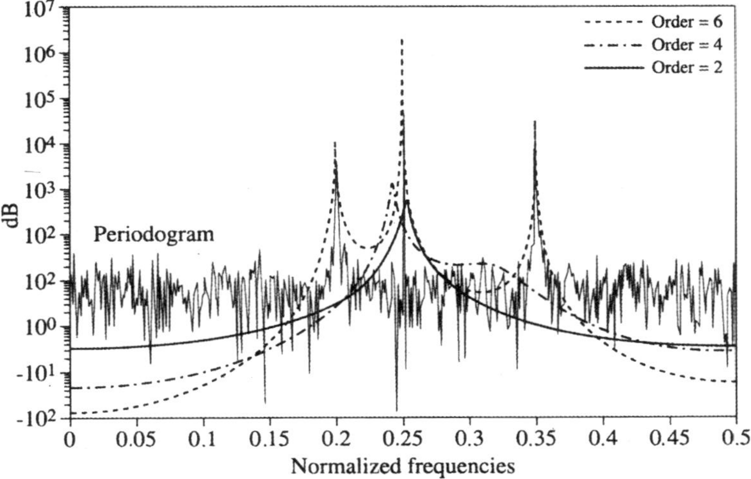

Figure 6.7. *AR spectrum evolution (compared to the periodogram) according to the order chosen for a signal made up of 3 sinusoids in white noise*

Figure 6.9 presents the results of Prony modeling on this same signal made up of three noisy sinusoids. Again, we realize that the three "lines" are better and better estimated. The interest of the Prony modeling is to provide a "physical interpretation" of the model by estimating the modal parameters of the analyzed signal in terms of frequencies, amplitudes, phases and dampings.

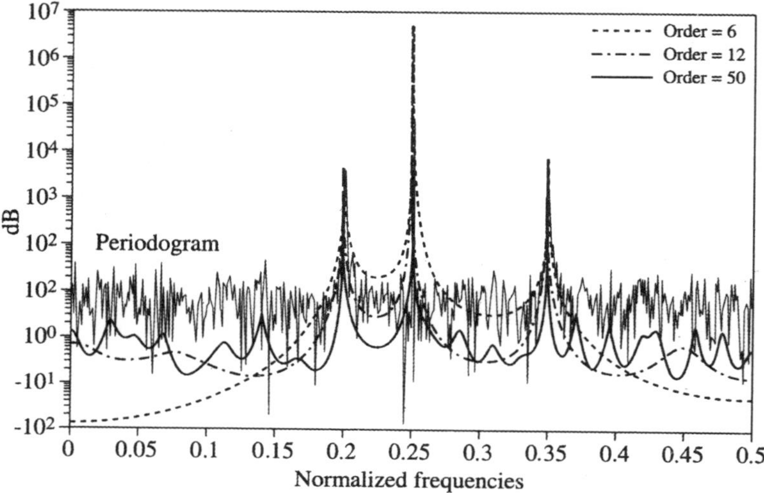

Figure 6.8. *AR spectrum evolution (compared to the periodogram) according to the order (from 6 to 50) chosen for a signal made up of 3 sinusoids in a white noise*

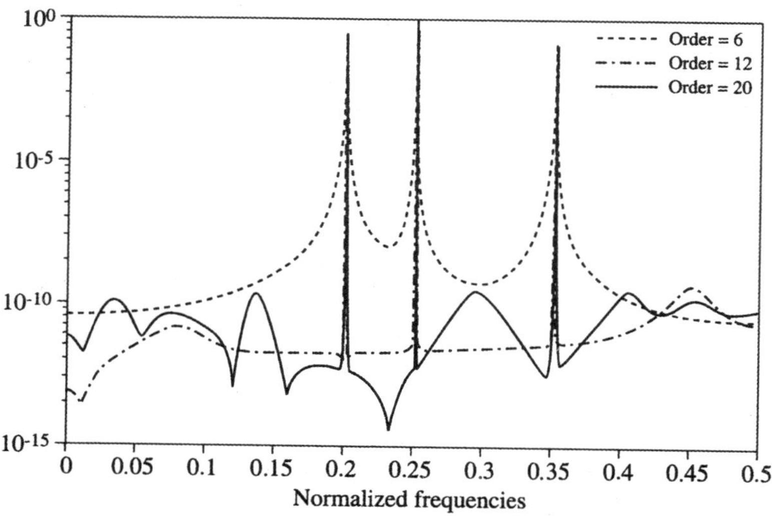

Figure 6.9. *Prony spectra at orders 6, 12 and 20 of the three noisy sinusoids*

The same type of illustration is performed on a real signal resulting from a vibrating structure. The MDL and AIC criteria applied on this signal indicate an AR modeling order between 17 and 21. The signal is short (of the order of 140 samples). An AR analysis (Figure 6.10) makes it possible to highlight a mode at the normalized frequency around 0.12 that the specialists of the field confirm.

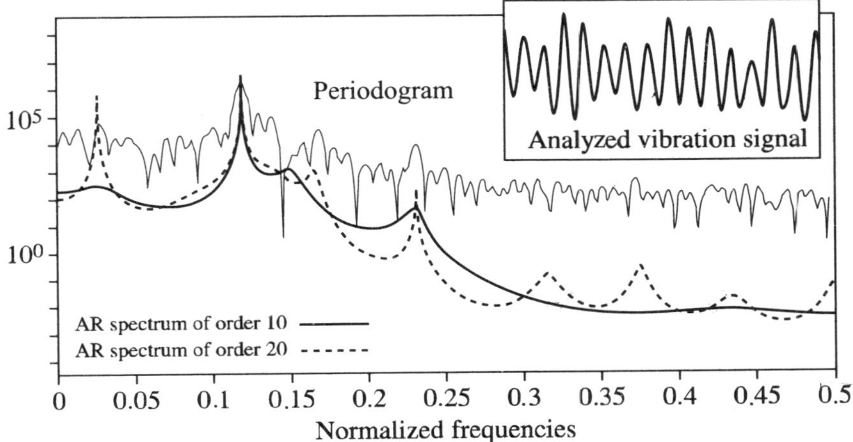

Figure 6.10. *AR modeling of a vibration signal*

This same mode is again better characterized using a Prony analysis (Figure 6.11), then making it possible to give all the physical quantities linked to this vibration node: frequency and amplitude in particular.

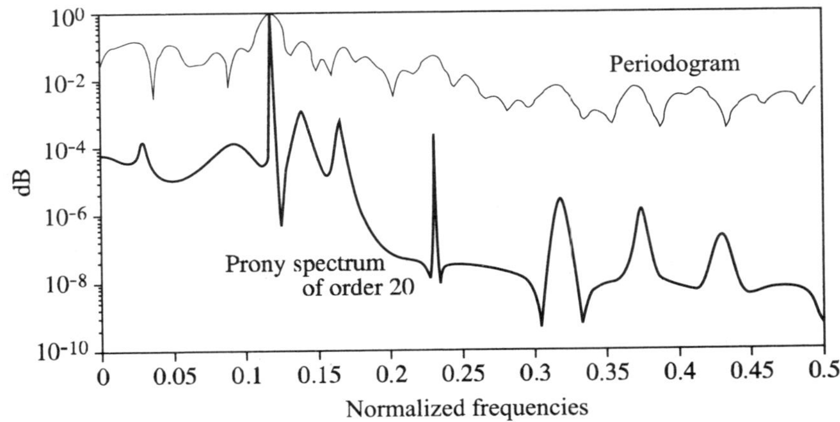

Figure 6.11. *Prony spectrum of the vibration signal*

6.4. Bibliography

[AKA 70] AKAIKE H., "Statistical predictor identification", *Ann. Insi. Statist. Math.*, vol. 22, p. 203-217, 1970.

[AKA 74] AKAIKE H., "A new look at the statistical model identification", *IEEE Trans. Autom. Contr.*, vol. 19, p. 716-723, December 1974.

[BRO 85] BROERSEN P., "Selecting the order of autoregressive models from small samples", *IEEE Trans. Acoust., Speech, Signal Processing*, vol. 33, no. 4, p. 874-879, August 1985.

[BUR 75] BURG J., Maximum entropy spectral analysis, PhD Thesis, Stanford University, Stanford, USA, 1975.

[BUR 85] BURSHTEIN D., WEINSTEIN E., "Some relations between the various criteria for autoregressive model order determination", *IEEE Trans. Acoust., Speech, Signal Processing*, vol. 33, no. 4, p. 1017-1019, August 1985.

[CHA 82] CHAN Y., LANGFORD R., "Spectral estimation via the HOYW equations", *IEEE Trans. Acoust., Speech, Signal Processing*, vol. 30, no. 5, p. 689-698, October 1982.

[DUC 95] DUCASSE A., MAILHES C., CASTANIÉ F., "Amplitude and phase estimator study in Prony method for noisy exponential data", *Proc. IEEE ICASSP-95*, Detroit, USA, p. 1796-1799, May 1995.

[DUC 97] DUCASSE A., Estimation de sous-harmoniques à l'aide de méthodes paramétriques, PhD Thesis, INP Toulouse, 1997.

[DUC 98] DUCASSE A., MAILHES C., CASTANIÉ F., "Estimation de fréquences: panorama des méthodes paramétriques", *Traitement du Signal*, vol. 15, no. 2, p. 149-162, 1998.

[FUC 88] FUCHS J.-J., "Estimating the number of sinusoids in additive white noise" *IEEE Trans. Acoust., Speech, Signal Processing*, vol. 36, no. 12, p. 1846-1853, December 1988.

[GIN 85] GINGRAS D., "Asymptotic properties of HOYW estimates of the AR parameters of a time series", *IEEE Trans. Acoust., Speech, Signal Processing*, vol. 33, no. 4, p. 1095-1101, October 1985.

[GOL 89] GOLUB G., VAN LOAN C., *Matrix Computations*, John Hopkins University Press, 1989.

[HAY 89] HAYKIN S., *Modern Filters*, Macmillan Publishing Company, 1989.

[KAS 80] KASHYAP R., "Inconsistency of the AIC rule for estimating the order of autoregressive models", *IEEE Trans. Autom. Contr*, vol. 25, no. 10, p. 996-998, 1980.

[KAY 80] KAY S. M., "Noise compensation for autoregressive spectral estimates", *IEEE Trans. Acoust., Speech, Signal Processing*, vol. 28, no. 3, p. 292-303, June 1980.

[KAY 88] KAY S. M., *Modern Spectral Estimation: Theory and Applications*, Prentice Hall, Englewood Cliffs (NJ), 1988.

[KON 88] KONSTANTINIDES K., YAO K., "Statistical analysis of effective singular values in matrix rank determination", *IEEE Trans. Acoust., Speech, Signal Processing*, vol. 36, no. 5, p. 757-763, May 1988.

[LAC 84] LACOUME J.-L. *et al.*, "Close frequency resolution by maximum entropy spectral estimators", *IEEE Trans. Acoust., Speech, Signal Processing*, vol. 32, no. 5, p. 977-983, October 1984.

[LAN 80] LANG S., MCCLELLAN J., "Frequency estimation with maximum entropy spectral estimators", *IEEE Trans. Acoust., Speech, Signal Processing*, vol. 28, no. 6, p. 720-789, December 1980.

[LEV 47] LEVINSON N., "The Wiener RMS error criterion in filter design and prediction", *J. Math. Phys.*, vol. 25, p. 261-278, 1947.

[MAK 77] MAKHOUL J., "Stable and efficient Lattice methods for linear prediction", *IEEE Trans. Acoust., Speech, Signal Processing*, vol. 25, no. 5, p. 423-428, October 1977.

[MAR 87] MARPLE S., *Digital Spectral Analysis with Applications*, Prentice Hall, Englewood Cliffs (NJ), 1987.

[MOR 77] MORF M., DICKINSON B., KAILATH T., VIEIRA A., "Efficient solution of covariance equations for linear prediction", *IEEE Trans. Acoust., Speech, Signal Processing*, vol. 25, no.5, p. 429-433, October 1977.

[PRO 92] PROAKIS J., RADER C., LING F., NIKIAS C., *Advanced Digital Signal Processing*, Macmillan Publishing Company, 1992.

[PUK 88] PUKKILA T., KRISHNAIAH P., "On the use of autoregressive order determination criteria in univariate white noise tests", *IEEE Trans. Acoust., Speech, Signal Processing*, vol. 36, no. 5, p. 764-774, May 1988.

[RIS 83] RISSANEN J., "A universal prior for the integers and estimation by minimum description length", *Ann. Stat.*, vol. 11, p. 417-431, 1983.

[SAK 79] SAKAI H., ARASE M., "Recursive parameter estimation of an autoregressive process disturbed by white noise", *Int. J.Contr*, vol. 30, p. 949-966, 1979.

[SÖD 91] SÖDERSTROM T., STOICA P., "On accuracy of HOYW methods for cisoids", *Proc. IEEE ICASSP-91*, p. 3573-3576, 1991.

[SÖD 93] SÖDERSTROM T., STOICA P., "Accuracy of High-Order Yule-Walker methods for frequency estimation of complex sine waves", *Proceedings IEE-F, Radar, Sonar and Navigation*, vol. 140, no.1, p. 71-80, February 1993.

[STO 89] STOICA P., SÖDERSTROM T., TI F., "Asymptotic properties of the high-order Yule-Walker estimates of sinusoidal frequencies", *IEEE Trans. Acoust., Speech, Signal Processing*, vol. 37, no.11, p. 1721-1734, November 1989.

[STO 92] STOICA P., SÖDERSTROM T., VAN HUFFEL S., "On SVD-based and TLS-based high-order Yule-Walker methods of frequency estimation", *Signal Processing*, vol. 29, p. 309-317, 1992.

[SWI 80] SWINGLER D., "Frequency errors in MEM processing", *IEEE Trans. Acoust., Speech, Signal Processing*, vol. 28, no. 2, p. 257-259, April 1980.

[VAN 91] VAN HUFFEL S., VANDEWALLE J., *The Total Least Squares Problem,* SIAM, 1991.

[WAX 85] WAX M., KAILATH T., "Detection of signals by information theoretic criteria", *IEEE Trans. Acoust., Speech, Signal Processing,* vol. 33, no. 2, p. 387-392, April 1985.

[WAX 88] WAX M., "Order selection for AR models by predictive least squares", *IEEE Trans. Acoust., Speech, Signal Processing*, vol. 36, no. 4, p. 581-588, April 1988.

[WIL 93] WILKES D., CADZOW J., "The effects of phase on high-resolution frequency estimators", *IEEE Trans. Signal Processing,* vol. 41, no. 3, p. 1319-1330, March 1993.

[YIN 87] YIN Y., KRISHNAIAH P., "On some nonparametric methods for detection of the number of signals", *IEEE Trans. Acoust., Speech, Signal Processing,* vol. 35, no. 11, p. 1533-1538, November 1987.

Chapter 7

Minimum Variance

In 1969 J. Capon proposed the following method which bears his name in the array processing domain:

> A high-resolution method of estimation is introduced which employs a wavenumber window whose shape changes and is a function of the wavenumber at which an estimate is obtained. It is shown that the wavenumber resolution of this method is considerably better than that of the conventional method [CAP 69].

Actually, he already presented this approach in a publication dating back to 1967 [CAP 67] where he mentioned Levin's previous publication [LEV 64]. In 1971, two years after J. Capon's publication, R. T. Lacoss adapts it to the time-series spectral analysis:

> The maximum likelihood method [...], which has been used previously (Capon, 1969) for wavenumber analysis with arrays, can be adapted to single time-series spectral analysis. One derivation of this is given below in terms of minimum variance unbiased estimators of the spectral components [LAC 71].

The method was initially called maximum likelihood. We will see why this name is not properly adapted (see section 7.4). The principle carried out is actually a variance minimization, hence its current name.

Chapter written by Nadine MARTIN.

Since then, the method has been used both in spectral and in spatial processing at the same time, the spatial processing being nevertheless privileged. Its interest is clearly expressed in J. Capon's introductory sentence: the window is adapted at each frequency. This adaptation faculty enables the Capon method to reach higher performance than the Fourier-based method. Let us illustrate it on one example. Figure 7.1 represents the spectral analysis of a noisy sinusoid by two approaches: the periodogram and the Capon method or minimum variance (MV).

This figure shows that the spectrum estimated by the MV method has a narrower main lobe and fewer secondary lobes, and this points out a gain in resolution and a weaker variance in relation to the periodogram. This gain is not without consequence. The method is digitally more intensive. But this point is not an obstacle nowadays and it will be less and less so in the future.

The MV method has an additional freedom degree in relation to those based on Fourier: the choice of a parameter called the order. This choice can be seen as a handicap. Actually, it is nothing. Contrary to the methods based on models such as the autoregressive model, the choice of this order does not have any link to the number of searched frequencies (see section 6.2.4). The higher the order is, the higher the resolution of the method will be. Moreover, while preserving better performances, it is possible to estimate a spectrum with a relatively weak order compared to the necessary window lengths in the Fourier methods. An order 6 for the MV analysis of a 16 sample simulated signal (bottom-right of Figure 7.1) is enough to correctly estimate the spectrum.

The spectral analysis of an acoustic signal provided by the DGA is represented in Figure 7.2. This signal with a duration of 16.6 s (50,000 samples, sampling frequency 3,000 Hz) is made up of an important set of sinusoids as well as of several structures with a wider spectral band. The MV method (top part of Figure 7.2) estimates the total of these structures with an order of 300, which means the inverse of a matrix 300×300. The interest of this approach, in spite of the calculation time, is then a drop-off in variance, which makes the extraction of various spectral structures easier. The periodogram (bottom part of Figure 7.2) also estimates the set of structures but the high variance requires the use of more complex techniques in order to extract the various spectral structures from it [DUR 99].

Another advantage of the MV approach is represented in Figure 7.3 with the spectral analysis of the same real signal as in Figure 7.2 but on a shorter duration of 43 ms (128 samples) instead of 16.6 s. A higher resolution makes it possible to estimate more spectral structures than a periodogram. An order of 80 is then enough.

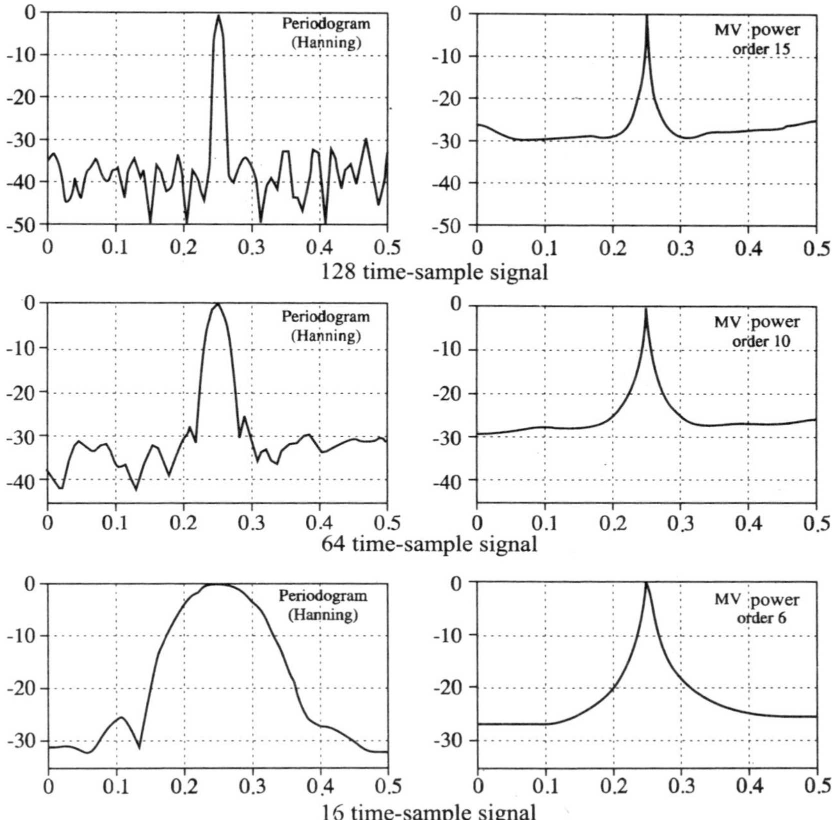

Figure 7.1. *Comparison of the simple periodogram weighted by a Hanning window (left column) and the minimum variance method (right column) for the estimation of the power spectral density of a sinusoid (0.25 Hz, sampling frequency 1 Hz), not very noisy (20 dB), on a time support of 128 samples, 64 samples and 16 samples. Horizontal axes: frequency. Vertical axes: spectrum in dB*

Experimental comparisons with other algorithms are developed on non-stationary signals in [BON 90, LAT 87]. Synthesis sheets were drawn up in order to briefly present a set of time-frequency approaches for the MV estimator [BAS 92]. See also [FER 86] for an adaptive MV estimator. As regards the applications, the reader will be able to consult [PAD 95, PAD 96] for an application in mechanics, [LEP 98] in seismics on the characterization of avalanche signals and [MAR 95] in room acoustics for an original adaptation in the time-octave domain.

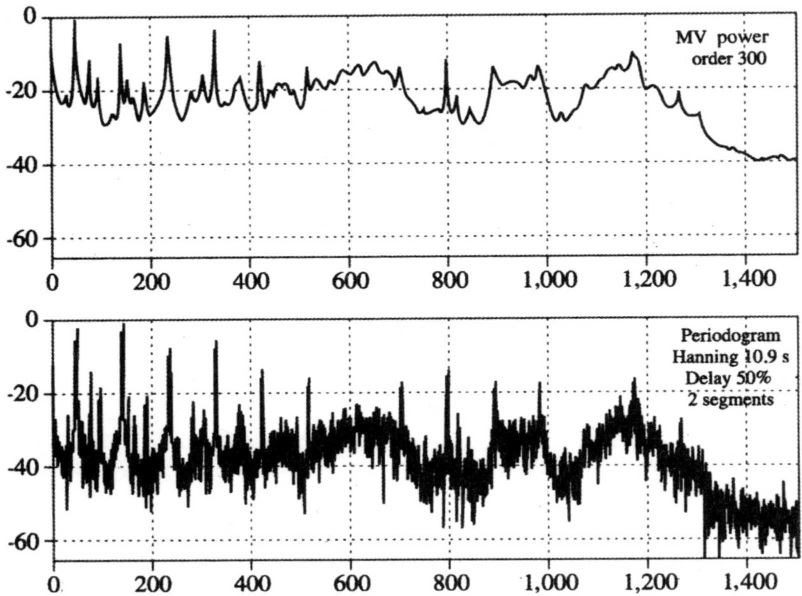

Figure 7.2. *Spectral analysis of an acoustic signal (duration 16.6 s, 50,000 samples, sampling frequency 3,000 Hz) by the MV method (top figure) and by the averaged periodogram method (bottom figure). Horizontal axes: frequency in Hz. Vertical axes: spectrum in dB*

In section 7.1, we present the MV method as a finite impulse response (FIR) filter bank. The method order, as we have just seen in Figures 7.1 and 7.2, corresponds to the number of coefficients of the filter impulse response, a filter being designed for each frequency channel. This concept of FIR filter bank places the MV method as an intermediate method between those based on Fourier and those based on models. The MV method designs a filter according to the analyzed signal while the Fourier methods are decomposed on a base of exponential functions, which are identical (see section 7.3).

The "parametric" term is often attributed to the MV method because the filter parameters need to be estimated (see section 7.1). However, it is important to note that these parameters do not contain all the spectral information contrary to the model-based methods. The MV method is considered as a high-resolution method that does not perform as well as the model-based method. Thus, this intermediate situation is due not only to the concept, but also to the properties obtained. In section 7.2, we insist on the particular structure of the filters designed this way, which lead to performances closely linked to the type of the analyzed signal. In this same section, we illustrate these performances by considering simulated signals of a relatively short duration on which the theoretical results established in an asymptotic context are not applicable.

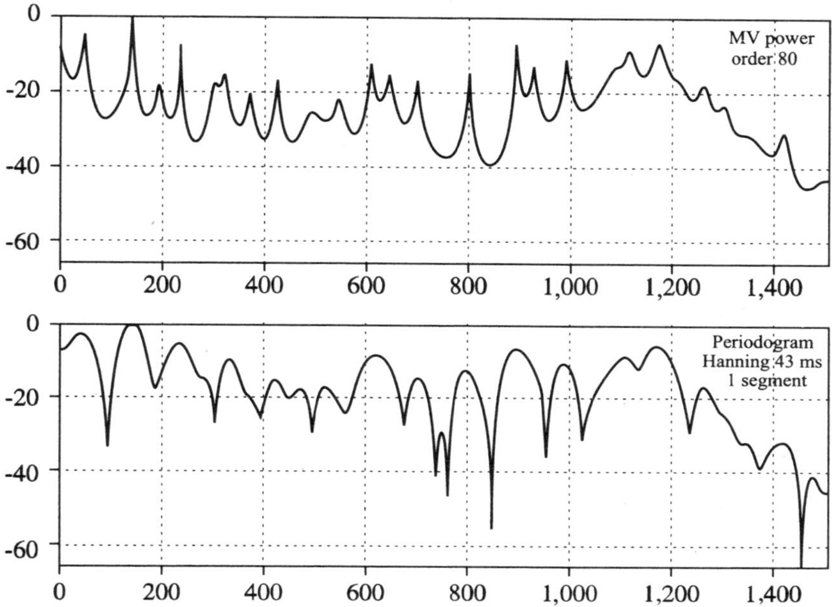

Figure 7.3. *Spectral analysis of an acoustics real signal (duration 43 ms, 128 samples, sampling frequency 3,000 Hz) by the MV method (top figure) and by the periodogram method (bottom figure). Horizontal axes: frequency in Hz. Vertical axes: spectrum in decibels*

Initiated by M.A. Lagunas, modifications were proposed and led to estimators called normalized MV. All these modifications are presented in section 7.5. More recently, the study of the properties of the MV filters led to a new estimator more adapted to the analysis of the mixed signals. This estimator, called CAPNORM, is described in section 7.6.

7.1. Principle of the MV method

Let $x(k)$ be a discrete, random, stationary signal sampled at frequency $f_s = 1/T_s$ and of power spectral density $S_x(f)$. The concept of the MV method, schematized in Figure 7.4, is based on the construction of a FIR filter at a frequency f_c, applied to the signal $x(k)$ according to the following two constraints:

– at frequency f_c, the filter transfers the input signal without modification;

– the filter output power of other frequencies than the frequency f_c is minimized.

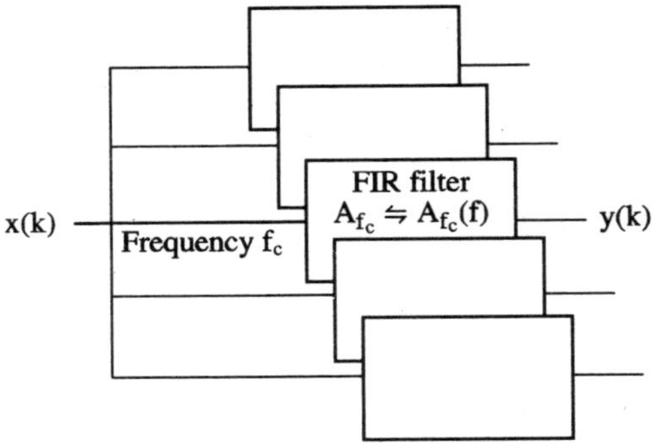

Figure 7.4. *Construction of a FIR filter bank for the analysis of the signal x(k) at all frequencies f_c between 0 and f_c /2*

The first constraint implies that the filter frequency response, noted $A_{f_c}(f_c)$, is equal to 1 at the frequency f_c, given:

$$A_{f_c}(f_c) = 1 \tag{7.1}$$

In order take the second constraint into account, the power P of the filter output $y(k)$ is expressed as the spectral density of the input using the relation:

$$P = E\left(y(k)^2\right) = \int_{-f_c/2}^{+f_c/2} \left|A_{f_c}(f)\right|^2 S_x(f)\,df \tag{7.2}$$

In order to set apart the frequency f_c in which we take interest, the power spectral density of the signal $x(k)$ is decomposed in two parts:

$$S_x(f) = S_x^o(f) + S_x^i(f)$$

with:

$$S_x^o(f) = \begin{cases} S_x(f) & \text{for } f_c - \varepsilon \le f \le f_c + \varepsilon \\ 0 & \text{otherwise} \end{cases}$$

$$S_x^i(f) = \begin{cases} S_x(f) & \text{for } f < f_c - \varepsilon \text{ and } f > f_c + \varepsilon \\ 0 & \text{otherwise} \end{cases}$$

knowing that $\varepsilon \ll f_c$.

The aim of this decomposition is to set apart the spectral information at frequency f_c. When ε tends to 0, the integral of the spectrum $S_x^o(f)$ tends to the searched quantity $P(f_c)$. By applying the first constraint defined by equation [7.1], when ε tends to 0, equation [7.2] becomes:

$$P = \int_{-f_c/2}^{+f_c/2} \left| A_{f_c}(f) \right|^2 S_x^i(f)\,df + P(f_c) \qquad [7.3]$$

with:

$$P(f_c) = \lim_{\varepsilon \to 0} \int_{-f_c/2}^{+f_c/2} S_x^o(f)\,df$$

The second constraint consists of minimizing the integral of equation [7.3]. Given the fact that the added term $P(f_c)$ is a constant, it is equivalent to minimizing the total output power P. By applying the Parseval relation, the dual expression is expressed according to the covariance matrix $\mathbf{R}_x$ of the input signal $x(k)$ and of the vector $\mathbf{A}_{f_c}$ of the coefficients of a FIR filter impulse response at frequency f_c:

$$P = \mathbf{A}_{f_c}^H \mathbf{R}_x \mathbf{A}_{f_c} \qquad [7.4]$$

$$A_{f_c}(f) = \mathbf{E}_{f_c}^H \mathbf{A}_{f_c} \qquad \mathbf{E}_f^T = \left(1, e^{2\pi jfTc}, \ldots, e^{2\pi j(M-1)fTc}\right)$$

with:

$$\mathbf{R}_x = E\left(\mathbf{X} \cdot \mathbf{X}^H\right) \text{ matrix } M \times M,\ \mathbf{X}^T = \left(x(k-M), \ldots, x(k-1)\right)$$

The exponents $(\ )^T$ and $(\ )^H$ represent transpose and hermitic transpose respectively.

The minimization of equation [7.4] under the constraint [7.1] is achieved by the technique of Lagrange multipliers [CAP 70]. It makes it possible to obtain the filter impulse response at frequency f_c:

$$\mathbf{A}_{f_c} = \frac{\mathbf{R}_x^{-1} \mathbf{E}_{f_c}}{\mathbf{E}_{f_c}^H \mathbf{R}_x^{-1} \mathbf{E}_{f_c}} \qquad [7.5]$$

and the output power P noted $P_{MV}(f_c)$:

$$P_{MV}(f_c) = \frac{1}{\mathbf{E}_{f_c}^H \mathbf{R}_x^{-1} \mathbf{E}_{f_c}} \qquad [7.6]$$

This power P is the Capon estimator, also called the minimum-variance estimator. The estimated quantity is linked to the searched power, the power $P(f_c)$ of the signal at frequency f_c, by equation [7.3]. Thus, it is clear that the more efficient the minimization is, the better the estimation is, i.e. the integral of equation [7.3] tends to 0.

Such a filter is built at frequencies ranged between 0 and $f_e/2$. The FIR bank filter defined this way does not require any particular hypothesis on the signal apart from the stationarity property and the existence of its power spectral density. The signal can be real or complex. No model is applied. The coefficients are automatically adapted starting from the signal autocorrelation matrix.

Indeed, in order to calculate equation [7.6], the autocorrelation matrix should be estimated. To this end, the reader will be able to refer to [GUE 85] and to Chapter 3 of this book. In addition, [HON 98b] and [JAN 99] consider for the Capon method the case of the progressive covariance matrices, estimated as in [7.4] starting from the data vector $\mathbf{X}^T = (x(k-M),\ldots,x(k-1))$, or retrograde, starting from $\mathbf{X}^T = (x(k+M),\ldots,x(k+1))$, or both progressive and retrograde at the same time. In this latter case, the statistic precision is increased, the signal being stationary.

It is interesting to note that the Capon estimator can be used as a covariance estimator for any other estimator of power spectral density simply by an inverse Fourier transform. Hongbin has shown the good statistic properties of this estimator [HON 98a].

7.2. Properties of the MV estimator

7.2.1. *Expressions of the MV filter*

In order to understand the internal mechanism of the MV estimator, it is very instructive to study the behavior of the designed filter by setting the constraints as explained in section 7.1. This filter has a shape, which depends on the signal; an analytical study can only be based on *a priori* given signals. A comprehensive conclusion can then be drawn as a result of an experimental study on more complex signals. In this section we present the main elements of such a study [DUR 00].

Let us consider the simple case where $x(k)$ is a complex exponential function at the frequency $f_{\exp}$ embedded in an additive, complex, centered white noise $b(k)$ of variance σ^2. Let $Ce^{j\phi}$ be the complex amplitude of this exponential, the signal is written in the following vector form:

$$\mathbf{X} = Ce^{j\phi}\mathbf{E}_{f_{\exp}} + \mathbf{B} \qquad [7.7]$$

where $\mathbf{X}$ and $\mathbf{E}_{f_{\exp}}$ are defined as in equation [7.4] and:

$$\mathbf{B}^T = \left(b(k-M),\ldots,b(k-1)\right)$$

The signal covariance matrix is written:

$$\mathbf{R}_x = \mathrm{C}^2\mathbf{E}_{f_{\exp}}\mathbf{E}_{f_{\exp}}^H + \sigma^2\mathbf{I} \qquad [7.8]$$

with $\mathbf{I}$ being the identity matrix of dimension M x M.

Using the Sherman-Morrison formula [GOL 85], we deduce the inverse of autocorrelation matrix:

$$\mathbf{R}_x^{-1} = \frac{1}{\sigma^2}\left(\mathbf{I} - \frac{\dfrac{\mathrm{C}^2}{\sigma^2}}{1+M\dfrac{\mathrm{C}^2}{\sigma^2}}\mathbf{E}_{f_{\exp}}\mathbf{E}_{f_{\exp}}^H\right) \qquad [7.9]$$

Let us assume:

$$Q = \frac{\dfrac{\mathrm{C}^2}{\sigma^2}}{1+M\dfrac{\mathrm{C}^2}{\sigma^2}} \qquad [7.10]$$

Let us note that $\dfrac{\mathrm{C}^2}{\sigma^2}$ represents the signal to noise ratio. By substituting equations [7.9] and [7.10] into equation [7.5], we obtain the expression of the MV filter impulse response:

$$\mathbf{A}_{f_c} = \frac{\mathbf{E}_{vc} - Q\mathbf{E}_{f_{\exp}}\mathbf{E}_{f_{\exp}}^H\mathbf{E}_{f_c}}{M - Q\left|\mathbf{E}_{f_c}^H\mathbf{E}_{f_{\exp}}\right|^2} \qquad [7.11]$$

We can deduce from this the frequency response of the filter adapted to the signal $x(k)$ at the frequency f_c:

$$A_{f_c}(f) = \frac{D(f_c - f) - QM\, D\left(f_{\exp} - f\right) D\left(f_c - f_{\exp}\right)}{1 - QM\left|D\left(f_c - f_{\exp}\right)\right|^2} \qquad [7.12]$$

with $D\left(f_k - f_i\right)$ the Dirichlet's kernel defined by:

$$D\left(f_k - f_i\right) = \frac{1}{M}\mathbf{E}_{f_i}^H\mathbf{E}_{f_k} = e^{\pi j(f_k - f_i)(M-1)T_e}\frac{\sin\left(\pi\left(f_k - f_i\right)MT_e\right)}{M\sin\left(\pi\left(f_k - f_i\right)T_e\right)} \qquad [7.13]$$

By substituting equation [7.9] into equation [7.6], we obtain the MV estimator expression:

$$P_{MV}\left(f_c\right) = \frac{\sigma^2}{M\left(1 - QM\left|D\left(f_c - f_{\exp}\right)\right|^2\right)} \qquad [7.14]$$

The function [7.12] is a relatively complex function which we will study for noticeable frequency values and for faraway signal to noise ratios.

If the filter frequency f_c is equal to the exponential frequency $f_{\exp}$, equations [7.12] and [7.14] are written:

$$A_{f_{\exp}}(f) = D\left(f_{\exp} - f\right) \qquad [7.15]$$

$$P_{MV}\left(f_c\right) = \frac{\sigma^2}{M(1 - QM)} = \frac{\sigma^2}{M} + C^2 \qquad [7.16]$$

The signal is filtered by a narrow band filter centered on the frequency of the exponential with a maximum equal to 1. In this particular case, the MV estimator is reduced to a Fourier estimator, a simple periodogram, non-weighted and non-averaged. The frequency response is a Dirichlet's kernel (see equation [7.15]). The output power given by equation [7.16] is dependent on the order M [LAC 71, SHE 91]. This dependence in $1/M$ is responsible for a bias, which diminishes when the order increases (see section 7.2.2).

For the other cases, it is necessary to make approximations. If the signal to noise ratio is weak ($C^2/\sigma^2 << 1$), Q tends to 0 and equations [7.12] and [7.14] are written:

$$A_{f_c}(f) \approx D(f_c - f) \quad [7.17]$$

$$P_{MV}(f_c) = \frac{\sigma^2}{M} \quad [7.18]$$

This drastic approximation also leads to a simple periodogram, non-weighted and non-averaged with, as for a frequency response, a Dirichlet's kernel centered on the MV filter frequency.

The other case is more interesting. It is about studying the case of the signal on noise bigger than $1(C^2/\sigma^2 >> 1)$. In this case, the product QM tends to 1 and the frequency response is no longer a Dirichlet's kernel. Equation [7.12] highlights the basic principle of the minimum variance estimator. While the denominator is constant, the numerator is the difference between two Dirichlet's kernels. The first is centered on the filter frequency f_c and the second is centered on the exponential frequency f_{exp}. This second kernel is also smoothed by a kernel which depends on the difference $(f_c - f_{exp})$ between these two frequencies. Let us examine the behavior of this impulse response according to the relative position of these two frequencies.

The filter frequency f_c is far from the frequency of the exponential f_{exp}

The smoothing factor $D(f_c - f_{exp})$ tends to 0. Then, equation [7.12] reduces to:

$$A_{f_c}(f) \approx \begin{cases} D(f_c - f) & \text{if } f \neq f_{exp} \\ 0 & \text{if } f \to f_{exp} \end{cases} \quad [7.19]$$

In this case, the response is a Dirichlet's kernel except at the frequency of the exponential. This latter point is the fundamental difference with a Fourier estimator.

The filter frequency f_c is close to the frequency of the exponential $f_{\exp}$

Given the fact that the term $(f_c - f_{\exp})$ is weak, the smoothing factor $D(f_c - f_{\exp})$, contrary to the preceding case, plays a predominant role. The impulse response of equation [7.12] generally has two predominant lobes. This response has a maximum which is greater than 1 and which is no longer at the filter frequency but at a frequency ranged between $f_{\exp}$ and f_c [DUR 00]. The filter is no longer a narrow band filter.

The preceding approximations were validated with simulations [DUR 00]. They lead to a key conclusion. The filter generated by the minimum variance method does not lead to narrow band filters, which was often claimed in the literature [KAY 88, LAG 86]. Actually, the constraints imposed by the minimum variance are not those of a narrow band filter. This characteristic has important consequences when an equivalent bandwidth should be calculated (see section 7.5.1).

Figure 7.5 illustrates the MV filters calculated when analyzing a mixed signal, sum of a narrow band filter and of a wide band filter. Figure 7.5 represented the signal power estimated by the MV method at order 12. It was necessary to calculate a frequency response at all frequencies of this spectrum. The other five figures are some examples of frequency responses at specific frequencies in relation to the studied signal. It is clear that only the frequency response calculated at the sinusoid frequency (Figure 7.5f) is similar to that of a narrow band filter. All the others (Figures 7.5b to 7.5e) have lobes whose maximal amplitude can reach 25, which is greatly superior to 1. Figures 7.5c and 7.5e are examples where, at the filter frequency, the lobe is not at its maximum.

7.2.2. *Probability density of the MV estimator*

Knowing the probability density of the MV estimator is the best way for accessing all the estimator moments. We will consider a more general signal than in the preceding paragraph. Let a complex signal $x(k)$ be the sum of an ordinary determinist signal $d(k)$ and of a Gaussian centered random noise $b(k)$. The vector $\mathbf{B}^T = \left(b(k-1),\ldots,b(k-M)\right)$ is normally distributed $\mathcal{N}\left(0,\mathbf{R}_b\right)$ where $\mathbf{R}_b = E\left(\mathbf{B}\cdot\mathbf{B}^H\right)$ represents the covariance matrix $M \times M$ of the noise $b(k)$. The vector $\mathbf{X}$ defined by $\mathbf{X}^T = (x(k-1),\ldots, x(k-M))$ is therefore normally distributed $\mathcal{N}\left(\mathbf{D},\mathbf{R}_b\right)$ where $D^T = \left(d(k-1),\ldots,d(k-M)\right)$.

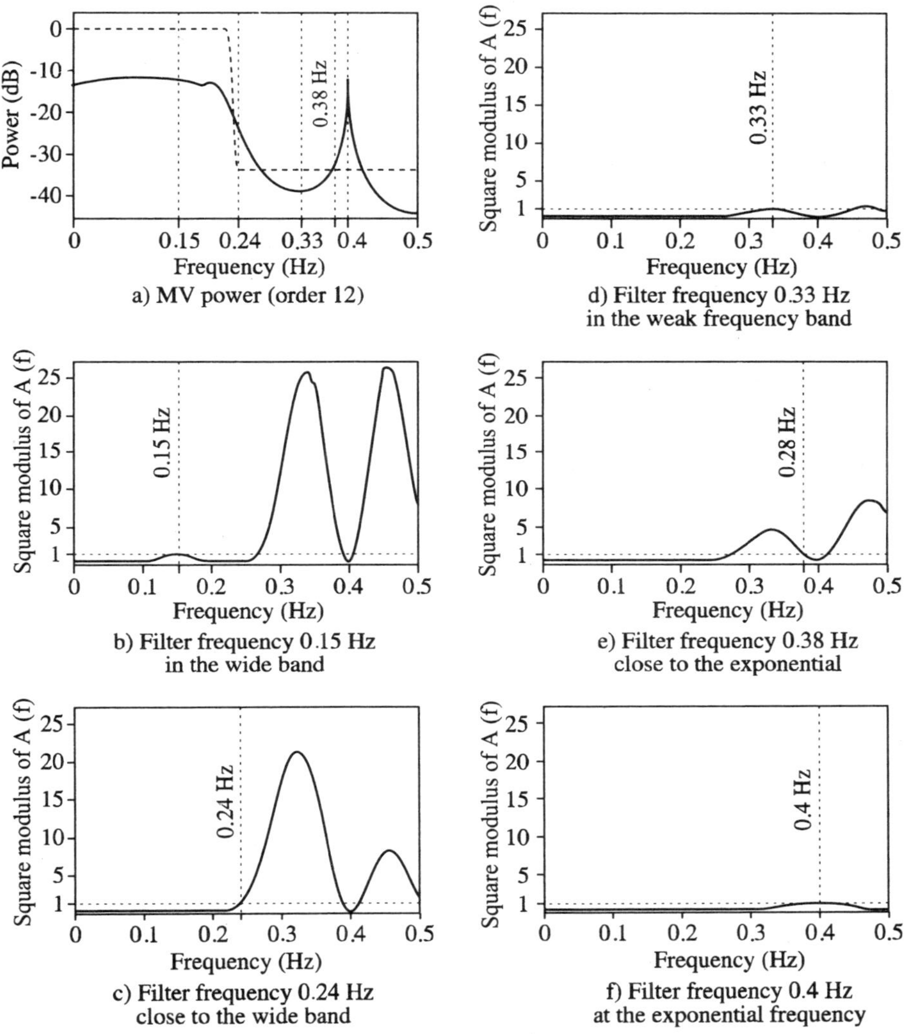

Figure 7.5. *Frequency response of the MV filters calculated starting from a simulated signal, sum of a sinusoid (0.4 Hz, amplitude 0.5, signal on noise ratio 25 dB) and of a white noise of variance 1 filtered by an elliptic low-pass filter (order 12, attenuation 80 dB from 0.22 Hz to 0.25 Hz), sampling frequency 1 Hz, 256 samples, (a) Signal power estimated by the MV method at the order* $M = 12$. *(b to f) Square modulus of the frequency responses of the MV filters at frequencies:* $f_c = 0.15$ *Hz (b),* $f_c = 0.24$ *Hz (c),* $f_c = 0.33$ *Hz(d),* $f_c = 0.38$ *Hz (e),* $f_c = 0.4$ *Hz (f)*

If the signal $x(k)$ is filtered by an MV filter defined by $\mathbf{A}_{f_c}$, the vector of the impulse response coefficients (see equation [7.5]), the filter output signal $y(k)$ is complex and is written:

$$y(k) = \mathbf{A}_{f_c}^T X \qquad [7.20]$$

The filtering operation being linear, the signal $y(k)$ is normally distributed:

$$\mathcal{N}\left(\mathbf{A}_{f_c}^T m_z, \mathbf{A}_{f_c}^H \mathbf{R}_b \mathbf{A}_{f_c}\right)$$

The square modulus of $y(k)$, $y(k)y^*(k)$ represents the Capon estimator according to equations [7.4] and [7.6] and is a non-central χ^2 with two degrees of freedom, of proportionality parameter $\mathbf{A}_{f_c}^H \mathbf{R}_d \mathbf{A}_{f_c}$ and of non-central parameter $\mathbf{A}_{f_c}^H \mathbf{R}_d \mathbf{A}_{f_c}$. This law represents the MV estimator probability density in a relatively general context because we did not make any particular hypothesis, either on the additive noise except for its Gaussianity or on the determinist signal, which can be made up of several components. The correlation matrices are supposed to be exact.

Knowing the law, we can deduce an expression of the mean and of the variance of the MV estimator:

$$\begin{aligned} E\left(P_{MV}(f_c)\right) &= \mathbf{A}_{f_c}^H \mathbf{R}_b \mathbf{A}_{f_c} + \mathbf{A}_{f_c}^H \mathbf{R}_d \mathbf{A}_{f_c} \\ Var\left(P_{MV}(f_c)\right) &= \left(\mathbf{A}_{f_c}^H \mathbf{R}_b \mathbf{A}_{f_c}\right)^2 + 2.\mathbf{A}_{f_c}^H \mathbf{R}_b \mathbf{A}_{f_c}.\mathbf{A}_{f_c}^H \mathbf{R}_d \mathbf{A}_{f_c} \end{aligned} \qquad [7.21]$$

It is interesting to note that these expressions are expressed in terms of the $\mathbf{A}_{f_c}^H \mathbf{R}_d \mathbf{A}_{f_c}$ and $\mathbf{A}_{f_c}^H \mathbf{R}_d \mathbf{A}_{f_c}$ quantities, which represent the output powers of the MV filter driving by the noise $b(k)$ alone or by the determinist signal $d(k)$ alone respectively, the filter being designed starting from the sum signal $x(k) = d(k) + b(k)$. These expressions are to be compared to those calculated for a periodogram (see section 7.3).

If the additive noise $b(k)$ is white of variance a^2, then $R_b = \sigma^2\mathbf{l}$, $\mathbf{I}$ being the identity matrix of dimension M x M. The mean and the variance of $P_{MV}(f_c)$ are written:

$$\begin{aligned} E\left(P_{MV}(f_c)\right) &= \sigma^2 \mathbf{A}_{f_c}^H \mathbf{A}_{f_c} + \mathbf{A}_{f_c}^H \mathbf{R}_d \mathbf{A}_{f_c} \\ Var\left(P_{MV}(f_c)\right) &= \sigma^4 \left(\mathbf{A}_{f_c}^H \mathbf{A}_{f_c}\right)^2 + 2\sigma^2.\mathbf{A}_{f_c}^H \mathbf{A}_{f_c}.\mathbf{A}_{f_c}^H \mathbf{R}_d \mathbf{A}_{f_c} \end{aligned} \qquad [7.22]$$

The mean expression, whether the noise is white or not, shows that the MV estimator is a biased estimator. The basic definition of the MV estimator summed up by equation [7.3] brings us to an identical conclusion, starting from the frequency domain:

$$Bias\left(P_{MV}\left(f_c\right)\right)=\int_{-f_c/2}^{+f_c/2}\left|A_{f_c}\left(f\right)\right|^2 S_x^i\left(f\right)df \qquad [7.23]$$

with $S_x^i\left(f\right)$ as defined in equation [7.2]:

$$S_x^i\left(f\right)=\begin{cases}S_x\left(f\right) & \text{for } f<f_c-\varepsilon \text{ and } f>f_c+\varepsilon\\ 0 & \text{else}\end{cases}$$

The MV estimator is built on the minimization of this integral which tends to 0 but which is never null. The MV estimator by construction can be only a biased estimator of the signal power. If we consider again the case of the noisy exponential function (see equation [7.7]), at the frequency of the exponential, the impulse response, the Fourier transform of the frequency response given by equation [7.15], is equal to the vector $\mathbf{E}_{f_{\exp}}$, hence the mean of the MV estimator is deduced from equation [7.21], which is now written:

$$E\left(P_{MV}\left(f_c\right)\right)=\frac{\sigma^2}{M}+C^2 \text{ for } f_c=f_{\exp} \qquad [7.24]$$

From this expression, the bias value is deduced:

$$Bias\left(P_{MV}\left(f_c\right)\right)=\frac{\sigma^2}{M} \text{ for } f_c=f_{\exp}$$

This expression is actually directly visible in equation [7.16]. This bias depends on the noise variance and especially on the order M. This formula illustrates the fact that the bias is not null, but it especially clearly shows the interest of a high order for diminishing the bias influence. These results are illustrated in Figure 7.6, which represents faraway cases. It is about the spectral analysis of a sinusoid embedded in a low-level noise (signal to noise ratio of 20 dB) at orders varying from 3 to 70. The representation is centered on the frequency of the sinusoid and the amplitudes are linearly represented in order to better visualize the bias. The amplitude of the spectrum minimum increases when M diminishes.

We will see in section 7.4 that, in order to have a non-biased estimator, it is necessary to possess *a priori* information which is stronger than that of the covariance matrix of the signal to be analyzed.

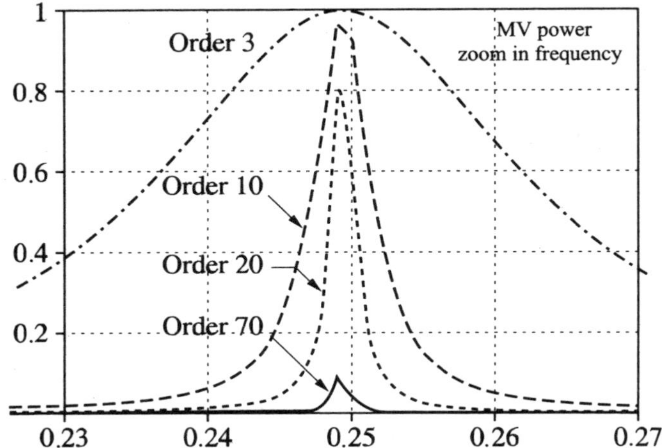

Figure 7.6. *Illustration of the bias and of the MV estimator resolution. Superposition of spectra estimated by the MV method at orders 3, 10, 20, and 70 on a signal made up of a noisy sinusoid (0.25 Hz, amplitude 1, sampling frequency 1 Hz, signal to noise ratio of 20 dB, 128 time samples). Horizontal axis: zoomed frequency around the frequency of the sinusoid. Vertical axis: linear spectrum*

As regards the variance given by equations [7.21] and [7.22], in the particular case of an exponential embedded in an additive white noise (see equation [7.7]), the variance is written at the exponential frequency:

$$\operatorname{var}\left(P_{MV}\left(f_c\right)\right) = \frac{\sigma^4}{M^2} + 2 \cdot \frac{\sigma^2 C^2}{M} \text{ for } f_c = f_{\exp}$$

This equation shows us again dependency of the variance in relation to the order M. Actually, it will be more interesting to calculate this variance at another frequency than the frequency of the exponential, or starting from the frequency response given by [7.12]. The calculations are more complex there and we will content ourselves with experimental observations.

[CAP 70] and [CAP 71] presented an experimental study in the array processing domain in order to argue the fact that the MV estimator has a weaker variance than a periodogram and an autoregressive estimator for long time-duration signals. This study was extended in spectral analysis in [LAC 71] and taken again in [KAY 88].

This weak variance is an important property of the MV estimator that we have already illustrated in Figures 7.1, 7.2 and 7.3. Figure 7.7 illustrates the variance diminution with the order for a sinusoid at a high signal to noise ratio of 20 dB. This figure clearly shows that for a high order relative to the signal size (order of 70 for 128 samples), the variance is of the same magnitude order as that of the periodogram. In Figure 7.2, where we present the analysis of a real signal with a high order (order of 300), the variance is still weak in relation to that of the periodogram because the signal size is of 50,000 samples. The effects of the estimation of the signal autocorrelation matrix are thus sensitive when the ratio of the order to the signal sample number is of the order of or more than 1/2.

For a complete asymptotic statistic study, the reader will be able to refer to the works of Liu and Sherman [LIU 98a] and to those of Ioannidis [IOA 94].

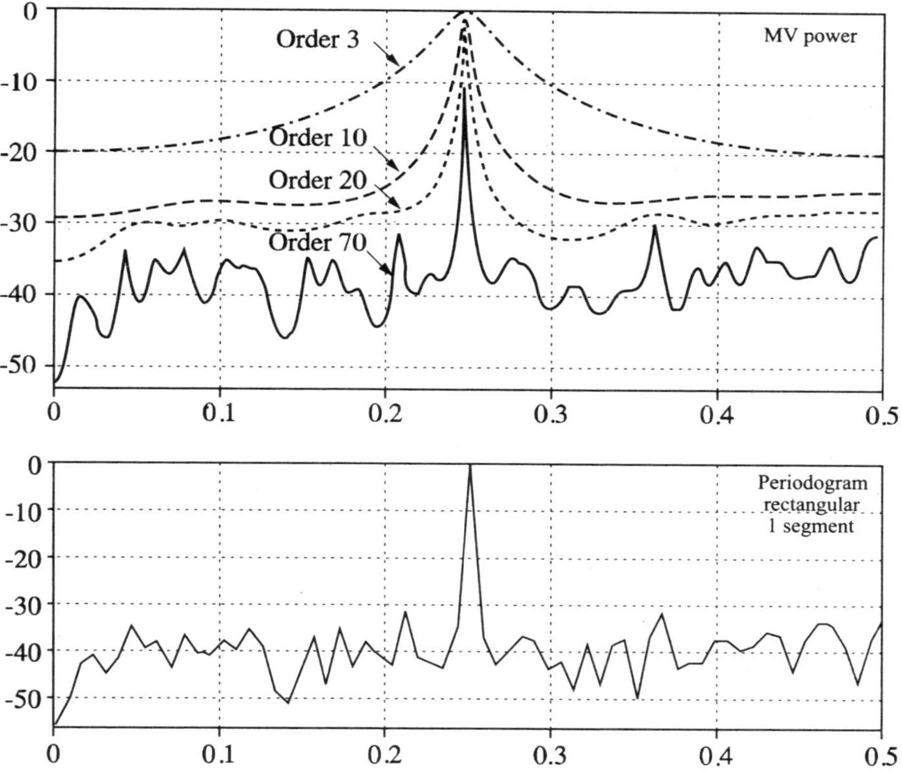

Figure 7.7. *Variance and resolution of the MV estimator for the analysis of a signal made up of a noisy sinusoid (0.25 Hz, sampling frequency 1 Hz, signal to noise ratio of 20 dB, time support 128 samples). Top: superposition of spectra estimated by the MV method at the orders 3, 10, 20, and 70. Bottom: simple non-weighted periodogram. Horizontal axes: frequency. Vertical axes: spectrum in dB*

7.2.3. *Frequency resolution of the MV estimator*

A partially analytical and partially numerical study is suggested in [LAC 71]. Lacoss qualifies the frequency resolution of a spectral estimator starting from the pass bandwidth at -3 dB, noted ΔB_{-3dB} and defined as a frequency band at half-maximum of a peak:

$$\Delta B_{-3dB} = f_+ - f_-$$

with $10\log(S(f_+)/S(f)) = 10\log(S(f_-)/S(f)) = -3$ for a peak at the frequency f of a spectrum $S(f)$.

Lacoss calculates an approximate expression of this band when the signal is made up of a noisy exponential as defined in equation [7.7] with a unit noise variance, given as $\sigma^2 = 1$, and for an exact covariance matrix defined for equation [7.8]. This study is performed at the same time on a non-weighted and non-averaged periodogram (parameters which induce the higher resolution), an MV estimator and an AR estimator. Lacoss obtains the following approximate expressions:

$$\left(\Delta B_{-3dB}\right)_{FOU} = \frac{\sqrt{6}}{\pi T_s M}$$

$$\left(\Delta B_{-3dB}\right)_{MV} = \left(\frac{3}{C^2 M}\right)^{1/2} \cdot \frac{2}{\pi T_s M}$$

$$\left(\Delta B_{-3dB}\right)_{AR} = \frac{2}{\pi T_s M^2 C^2}$$

with $\left(\Delta B_{-3dB}\right)_{FOU}$, the pass bandwidth at $-3dB$ of the simple periodogram (see equation [7.25]), $\left(\Delta B_{-3dB}\right)_{MV}$, that of the MV estimator (see equation [7.14]), and $\left(\Delta B_{-3dB}\right)_{AR}$ that of the AR estimator [LAC 71].

Starting from these equations, with a given M, an MV estimator has a better resolution than a periodogram but worse than an autoregressive estimator. Lacoss also highlights the nature difference between these estimators, the MV estimator being a power estimator and not one of power spectral density. This property that we will also encounter again in section 7.5 is fundamental for the comparison of the estimators.

Figure 7.7 illustrates the evolution of the MV method resolution with the order of the filters for a not very noisy sinusoid (signal on noise ratio of 20 dB); Figure 7.6

shows a zoom of it at the central frequency. It is clearly in evidence that the quantity ΔB_{-3dB} of the estimated peak diminishes with the order, hence an improvement of the frequency resolution with the order of the MV estimator.

It is interesting to know the capacity of a spectral estimator to separate two close frequencies. Figure 7.8 illustrates this property for a signal whose noise to ratio signal is of 10 dB. The three cases that were presented are compared to the periodogram in its more resolutive version, that is, non-averaged and non-weighted. Two peaks are separated if the amplitudes of the emerging peaks are higher than the estimator standard deviation. Thus, in the top part of Figure 7.8, the difference between the two frequencies is 0.02 Hz in reduced frequency and the MV estimator as well as the periodogram separate the two peaks. In the middle of Figure 7.8, the difference between the two frequencies is only 0.01 Hz. The MV estimator still separates the two peaks while the periodogram presents only one peak. However, a more detailed study of the periodogram could conclude on the possible presence of two non-resolved peaks by comparing the obtained peak with the spectral window of the estimator. Actually, the peak obtained by the periodogram has a bandwidth greater than the frequency resolution of the periodogram (bandwidth at -3 dB for 128 samples equal to 0.008 Hz) [DUR 99]. In Figure 7.8 at the bottom, where the frequency difference is only of 0.006 Hz, the estimators do not separate the peaks even if two maxima are observed on the MV estimator. These maxima do not emerge relatively to the estimator standard deviation.

7.3. Link with the Fourier estimators

The comparison with the Fourier-based methods is immediate if the estimator is written in the form of a filtering. A periodogram or a non-averaged correlogram $S_{FOU}(f)$ can be written:

$$S_{FOU}(f) = \frac{1}{M}\mathbf{E}_f^H \mathbf{R}_x \mathbf{E}_f \qquad [7.25]$$

This result is well known. The Fourier estimator is the output power of a filter bank whose impulse response of each filter at the frequency f is an exponential function at this frequency, a function limited in time by a rectangular window of M samples, the length of the vector $\mathbf{E}_f$. This vector is the counterpart of $\mathbf{A}_{f_c}$ for the Capon method (see equation [7.4]) but is totally independent of the signal. The power is divided by $1/M$ so that the result is the counterpart to a spectrum (see section 7.5.1).

The statistic performances of a periodogram are well known. We transcribe them in this section by using the same notations as the statistic study of the MV estimator in section 7.2.2. The periodogram $S_{FOU}(f)$ is a non-central χ^2, with two degrees of freedom, having the proportionality parameter $\mathbf{E}_{f_c}^H \mathbf{R}_b \mathbf{E}_{f_c}$ and non-central parameter $\mathbf{E}_{f_c}^H \mathbf{R}_d \mathbf{E}_{f_c}$, hence the expressions of the mean and of the variance

$$E\left(S_{FOU}(f)\right) = \frac{1}{M}\left(\mathbf{E}_f^H \mathbf{R}_b \mathbf{E}_f + \mathbf{E}_f^H \mathbf{R}_d \mathbf{E}_f\right)$$

$$Var\left(S_{FOU}(f)\right) = \frac{1}{M^2}\left(\left(\mathbf{E}_f^H \mathbf{R}_b \mathbf{E}_f\right)^2 + 2 \cdot \mathbf{E}_f^H \mathbf{R}_b \mathbf{E}_f \cdot \mathbf{E}_f^H \mathbf{R}_d \mathbf{E}_f\right)$$

which should be directly compared with [7.21]:

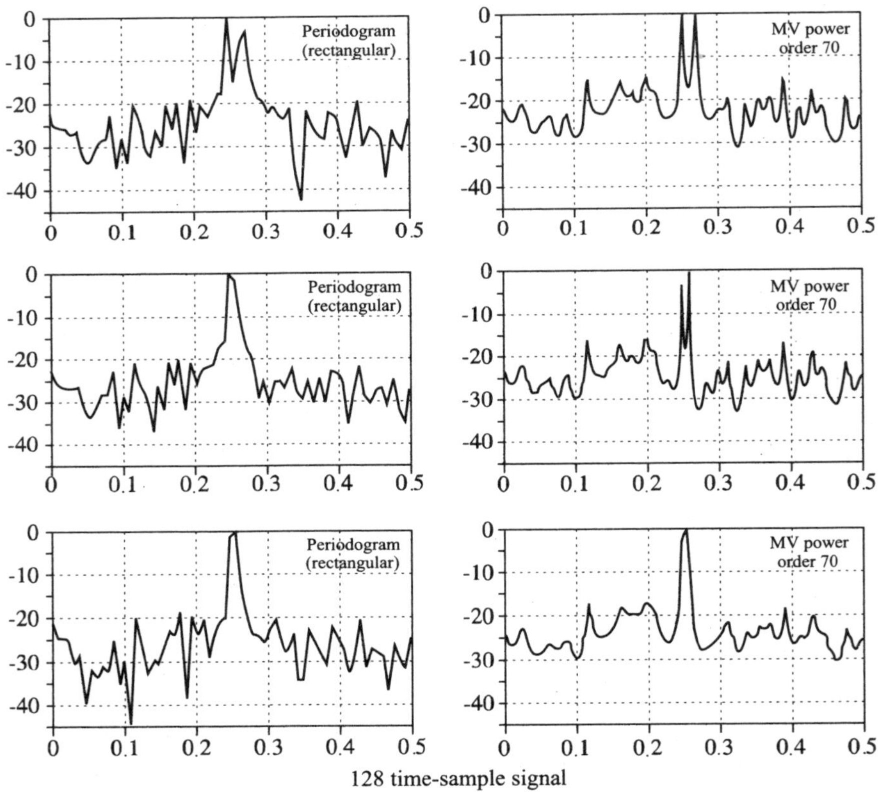

Figure 7.8. *Spectral analysis of the sum of two sinusoids of the same amplitude 1, of sampling frequency 1 Hz, of signal to noise ratio of 10 dB, on a time support of 128 samples by the MV estimator (order 70) (right column) comparatively to the periodogram (non-averaged and non-weighted) (left column). Frequencies of the sinusoids: 0.25 and 0.27 Hz (at the top), 0.25 and 0.26 Hz (in the middle), 0.25 and 0.256 Hz (at the bottom). Horizontal axes: frequency. Vertical axes: spectrum in dB*

In order to make the link with the more classic scalar expressions, we should note that $\frac{1}{M}\mathbf{E}_f^H\mathbf{R}_d\mathbf{E}_f$ represents the periodogram of the determinist signal and that, in the case of a white noise, the expression $\mathbf{E}_f^H\mathbf{E}_f$ appears, which is equal to *M*.

Let us continue the reasoning with a smoothed periodogram $\tilde{S}_{FOU}(f)$:

$$\tilde{S}_{FOU}(f)=\frac{1}{M}\mathbf{E}_f^H\tilde{\mathbf{R}}_x\mathbf{E}_f \qquad [7.26]$$

where the covariance matrix $\tilde{\mathbf{R}}_x$ is calculated starting from the signal *x*(*k*) weighted by a window. Let $\mathbf{G}_{FOU}$ be a diagonal matrix $M \times M$ whose elements are the samples of this weighting window; the matrix $\tilde{\mathbf{R}}_x$ is written:

$$\tilde{\mathbf{R}}_x=E\left((\mathbf{G}_{FOU}\mathbf{X})(\mathbf{G}_{FOU}\mathbf{X})^H\right)(\mathbf{G}_{FOU}\mathbf{X})(\mathbf{G}_{FOU}\mathbf{X})^H \qquad [7.27]$$

Thus, equation [7.25) becomes:

$$\tilde{S}_{FOU}(f)=\frac{1}{M}\left(\mathbf{E}_f^H\mathbf{G}_{FOU}\right)\mathbf{R}_x\left(\mathbf{G}_{FOU}{}^H\mathbf{E}_f\right) \qquad [7.28]$$

If this latest expression is compared to the equation of the MV method obtained by combining equations [7.4] and [7.5], we note that the MV method consists of applying on the signal a particular weighting window induced by the method and defined by the matrix $\mathbf{G}_{MV}$ which is written:

$$\mathbf{G}_{MV}=\frac{\mathbf{R}_x^{-1}}{\mathbf{E}_f^H\mathbf{R}_x^{-1}\mathbf{E}_f} \qquad [7.29]$$

This matrix is no longer diagonal as in the Fourier estimators. It is a function of the signal and of the frequency. $\mathbf{G}_{MV}$ is equal to $\mathbf{R}_x^{-1}$, the inverse of the covariance matrix of the analyzed signal, weighted by a quadratic form function of the frequency. This quadratic form $\mathbf{E}_f^H\mathbf{R}_x^{-1}\mathbf{E}_f$ is actually the power of a signal, which would have for covariance matrix $\mathbf{R}_x^{-1}$, this factor can thus be seen as a power normalization factor.

7.4. Link with a maximum likelihood estimator

By comparing the form of the equations, Capon [CAP 69] and Lacoss [LAC 71] concluded that the minimum variance method was equivalent to a maximum likelihood method. Since 1977, Lacoss recognized the "strange" meaning of this terminology, but he wrote that it is "too late" to change it [LAC 77]. Eleven years later, Kay examined the difference between these two estimators again [KAY 88].

Let us consider the simple case where the signal $x(k)$ is a complex exponential function at the frequency f_{exp} embedded in an additive complex white, Gaussian, centered noise $b(k)$ of variance σ^2. Let C_c be the complex amplitude of this exponential, the signal is described in the following vector form:

$$\mathbf{X} = C_c \mathbf{E}_{f_{exp}} + \mathbf{B} \qquad [7.30]$$

where $\mathbf{X}, \mathbf{E}_{f_{exp}}$ and $\mathbf{B}$ are defined in equation [7.7].

If the frequency f_{exp} is supposed to be known, it is possible to calculate an estimation $\hat{C}_c$ of the complex amplitude C_c by the maximum likelihood method. Because the noise $b(k)$ is Gaussian, maximizing the probability density of $\left(\mathbf{X} - C_c \mathbf{E}_{f_{exp}}\right)$ means minimizing the quantity $\left(\mathbf{X} - C_c \mathbf{E}_{f_{exp}}\right)^H \mathbf{R}_b^{-1} \left(\mathbf{X} - C_c \mathbf{E}_{f_{exp}}\right)$ with $\mathbf{R}_b$ the noise correlation matrix. The minimization leads to the following result [KAY 88]:

$$\hat{C}_c = \mathbf{H}^H \mathbf{X} \text{ with } \mathbf{H} = \frac{\mathbf{R}_b^{-1} \mathbf{E}_{f_{exp}}}{\mathbf{E}_{f_{exp}}^H \mathbf{R}_b^{-1} \mathbf{E}_{f_{exp}}} \qquad [7.31]$$

The bias and the variance of C_c are written:

$$E\left(\hat{C}_c\right) = \mathbf{H}^H C_c \mathbf{E}_{f_{exp}} = C_c \qquad [7.32]$$

$$\operatorname{var}\left(\hat{C}_c\right) = E\left(\left(\hat{C}_c - E\left(\hat{C}_c\right)^2\right)\right) = \mathbf{H}^H \mathbf{R}_b \mathbf{H} = \frac{1}{\mathbf{E}_{f_{exp}}^H \mathbf{R}_b^{-1} \mathbf{E}_{f_{exp}}} \qquad [7.33]$$

The maximum likelihood estimator can be interpreted as the output of a filter whose input is the signal $x(k)$ and of impulse response $\mathbf{H}$. Another way of proceeding leads to the same expression of the estimator $\hat{C}_c$. It is about minimizing the filter output variance given by the first part of equation [7.33] by constraining the bias to zero. The bias expression in [7.32] makes it possible to express the constraint in the form:

$$\mathbf{H}^{H}\mathbf{E}_{f_{\exp}} = 1 \qquad [7.34]$$

Thus, the maximum likelihood estimator is also a minimum variance estimator but is non-biased.

Comparing this approach with that of the Capon method brings the following conclusions. The constraint imposed on the two filters is the same (see equations [7.1] and [7.34]), but the likelihood stops there. Figure 7.9 illustrates the fact that the maximum likelihood estimator is the filtering output, while the Capon estimator is the output of a filtering followed by a quadration. The nature of estimations is thus very different. Capon is homogenous to a power or variance, while the maximum likelihood estimator is homogenous to an amplitude.

As the maximum likelihood estimator has a maximal variance, in the two cases, it is the filter output variance that is minimized. But this minimization is not put in the same terms, which leads to different impulse responses: that of the Capon method depends on the signal covariance matrix (see equation [7.4]) while that of the maximum likelihood estimator depends on the noise covariance matrix (see equation [7.31]).

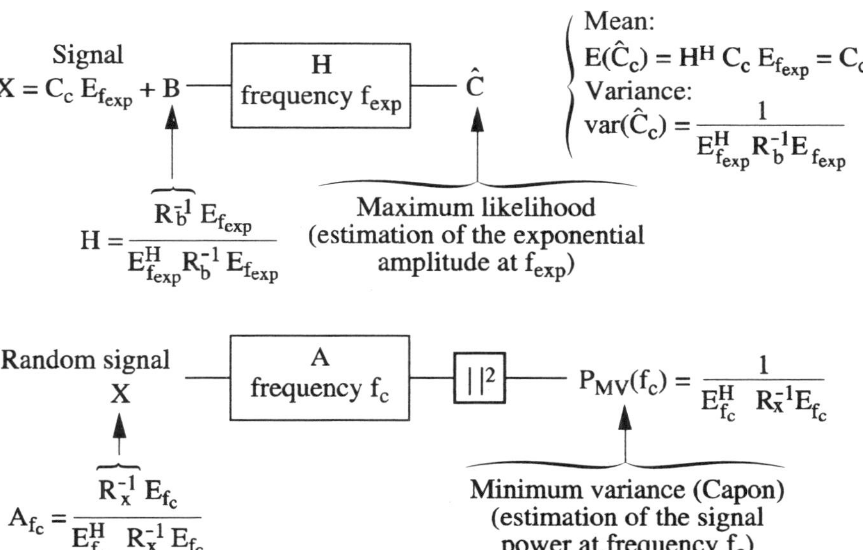

Figure 7.9. *Minimum variance and maximum likelihood*

When the analyzed signal has more than one exponential, the maximum likelihood estimator of the amplitude of one of the exponentials makes it necessary to know the added noise covariance matrix of the other exponentials.

These filter structures imply that the bias of the maximum likelihood estimator is null while the Capon estimator is biased (see section 7.2.2).

Thus, the maximum likelihood estimator is more efficient because it is non-biased, but it requires a strong *a priori* knowledge, which is not always available. As regards the approximate maximum likelihood estimators, the reader will be able to refer to the methods developed by P. Stoica called MAFI for "MAtched FIlter bank" and APES for "Amplitude and Phase Estimation of a Sinusoid" [LI 98, LIU 98b, STO 98, STO 99]. A recent approach combines the APES estimator and the Capon estimator in order to compensate the bias of the first one in frequency and of the second one in amplitude [JAK 00].

The Capon estimator is thus more general. It applies to any random signal. The Capon method does not estimate the amplitude of an exponential, but of the signal power at a given frequency. The filter impulse response is calculated starting from the signal covariance matrix which we can always estimate starting from data. The price to pay is a bias that is no longer null.

7.5. Lagunas methods: normalized and generalized MV

7.5.1. *Principle of normalized MV*

The output power $P_{MV}(f_c)$ of an MV filter is homogenous to a power and not to a power spectral density. M. A. Lagunas [LAG 84, LAG 86] proposed a modification of the method so that the area under the modified estimator represents the total power of the analyzed signal.

Using the hypothesis that the true spectral density $S_x(f)$ of the signal $x(k)$ is constant around the frequency f_c and equal to $S_x(f_c)$, equation [7.2] which represents $P_{MV}(f_c)$ can be written:

$$P_{MV}(f_c) \approx S_x(f_c) \int_{-f_c/2}^{+f_c/2} \left|A_{f_c}(f)\right|^2 df \qquad [7.35]$$

Constraint [7.1] being verified, the power $P_{MV}(f_c)$ is linked to the density S_x (f_c) by a factor that we will note as $\beta_{NMV}(f_c)$. By applying the Parseval relation, this factor can be written:

$$\beta_{NMV}(f_c) = \int_{-f_c/2}^{+f_c/2} \left|A_{f_c}(f)\right|^2 df = \frac{1}{T_s}\mathbf{A}_{f_c}^H \mathbf{A}_{f_c} \qquad [7.36]$$

By combining equations [7.35] and [7.36], M. A. Lagunas proposed a normalized estimator that we will note as $SNMV(f_c)$ and which is written:

$$S_{NMV}(f_c) = \frac{P_{MV}(f_c)}{\beta_{NMV}} \qquad [7.37]$$

This estimator can be expressed according to the impulse response of the Capon filter by considering equations [7.4] and [7.37]; hence:

$$S_{NMV}(f_c) = T_s \frac{\mathbf{A}_{f_c}^H \mathbf{R}_x \mathbf{A}_{f_c}}{\mathbf{A}_{f_c}^H \mathbf{A}_{f_c}} \qquad [7.38]$$

Substituting $\mathbf{A}f_c$ from equation [7.5] into equation [7.35] gives the expression of the NMV estimator:

$$S_{NMV}(f_c) = T_s \frac{\mathbf{E}_{f_c}^H \mathbf{R}_x^{-1} \mathbf{E}_{f_c}}{\mathbf{E}_{f_c}^H \mathbf{R}_x^{-2} \mathbf{E}_{f_c}} \qquad [7.39]$$

According to its definition (see equation [7.36]) and knowing that the filter frequency response equals 1 at the frequency f_c, the coefficient $\beta_{NMV}(f_c)$ has wrongly been assimilated to the equivalent filter band [LAG 86]. Actually, this name makes the implicit hypothesis that the frequency responses of the MV filters are null outside the vicinity of f_c. The MV filter study presented in section 7.2.3 shows that these filters are not narrow band filters. Other very important lobes are generally present. Thus, for most filters, the maximum of the frequency response is not at the frequency f_c and its value is greatly superior to 1. Therefore, the definition of equation [7.36] does not absolutely correspond to that of an equivalent band, even in magnitude order.

Moreover, this implicit hypothesis of the narrowband filters, which is necessary in order to write equation [7.35], is detrimental to the properties of this estimator. This consequence is discussed in section 7.5.2.

Most frequently, the proposed normalization [IOA 94, KAY 88] consists of dividing the power $P_{MV}(f_c)$ by the quantity:

$$B_e = \frac{1}{T_s}\mathbf{E}_{f_c}^H \mathbf{E}_{f_c} = \frac{1}{T_s M} \qquad [7.40]$$

This quantity B_e is the equivalent bandwidth of the impulse response $\mathbf{E}_{f_c}$ of a Fourier estimator and not that of the response $\mathbf{A}_{f_c}$ of a MV filter. It is thus about an approximation. The statistic performances of this estimator are studied in [IOA 94].

An alternative normalization is proposed in section 7.6.

7.5.2. *Spectral refinement of the NMV estimator*

The NMV method is not designed to have a high resolution. Its objective is to estimate a homogenous quantity to a power spectral density and not to a power. The frequency resolution is not improved, but the visualization makes it possible to better separate close frequencies. In Figure 7.10 we present the improvement brought by the NMV estimator in relation to the MV estimator when two frequencies are close. At a finite order M, the peaks sharpen, but they still remain centered on the same frequency.

This effect is due to the normalization by the factor $\beta_{NMV}(f_c)$ and it can be explained starting from the MV filter study presented in section 7.2 ([DUR 00]). Let us calculate the ratio $\dfrac{P_{MV}(f)}{S_{NMV}(f)}$ at the exponential frequency $f_{\exp}$ and at a close frequency ($f_{\exp} + \varepsilon$) with ε tending to 0.

When the filter frequency f_c is equal to the frequency $f_{\exp}$, the frequency response is given by equation [7.15]. In this case, the normalization factor defined by equation [7.36] and which we will note $\beta_{NMV_{\exp}}(f_c)$ is written:

$$\beta_{NMV_{\exp}}(f_c) \approx \int_{-f_c/2}^{+f_c/2} \left|D(f_{\exp} - f)\right|^2 df \qquad [7.41]$$

When the filter frequency f_c is equal to the frequency ($f_{\exp}$ + ε), the frequency response is given by equation [7.12] that we can approximate at order 2 [DUR 00]. The normalization factor, the normalization noted by $\beta_{NMV_e}(f_c)$ in this case, becomes:

$$\beta_{NMV_x}(f_c) = \int_{-f_c/2}^{+f_c/2} \left| \frac{D(f_{\exp} + \varepsilon - f) - D(f_{\exp} - f)\left(1 - (\pi M T_s)^2 \varepsilon^2 / 6\right)}{(\pi M T_s)^2 \varepsilon^2 / 3} \right|^2 df \qquad [7.42]$$

When ε tends to 0, we can show that:

$$\beta_{e\exp}(f_c) \ll B_{e\varepsilon}(f_c) \qquad [7.43]$$

This important relation shows that this factor considerably increases as soon as the frequency moves away from the peak maximum. This relation makes it possible to obtain the following order relation between the amplitude ratios:

$$\frac{S_{NMV}(f+\varepsilon)}{P_{MV}(f+\varepsilon)} \ll \frac{S_{NMV}(f)}{P_{MV}(f)} \qquad [7.44]$$

Close to a spectrum maximum, the NMV estimator sharpens its peak without keeping the amplitudes. The spectrum is thus subjected to a transformation which accentuates each spectral variation and which, with a finite order M, does not correctly estimate the spectrum amplitudes. This property is illustrated in Figure 7.11, which presents the NMV spectrum of the same signal as Figure 7.5. In the wideband part, the spectrum is estimated with an amplitude which is much lower than the true spectrum and with unrealistic oscillations. On the contrary, the brutal variation of the normalization factor around the pure frequency explains the improvement of this estimator at this frequency in relation to an MV estimator (see Figure 7.5).

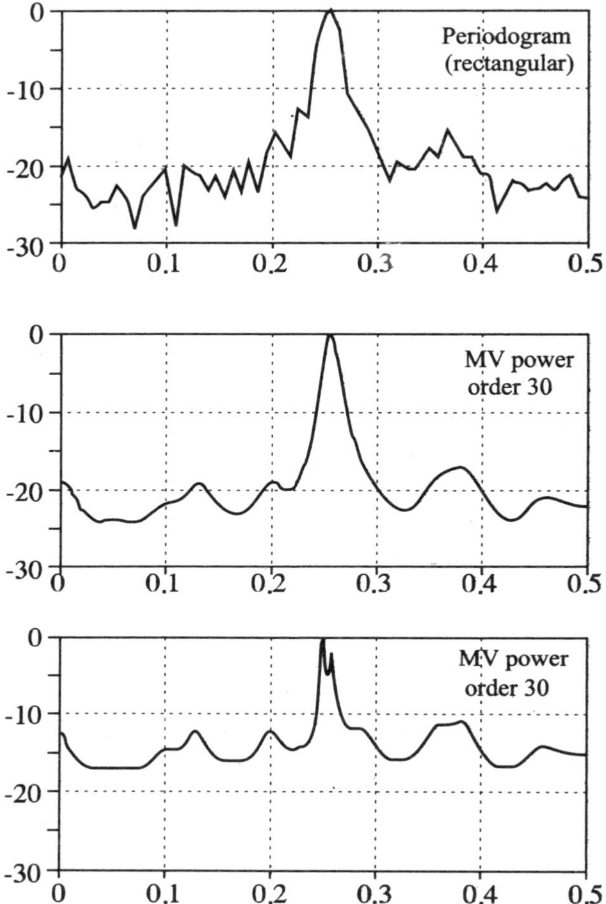

Figure 7.10. *NMV estimator in relation to the MV estimator and to the periodogram for the estimation of the power spectral density of the sum of two sinusoids at close frequencies (0.25 and 0.26 Hz, sampling frequency 128 time-sample signal sampling at 1 Hz) of the same amplitude 1 and of the signal on noise ratio 10 dB. Horizontal axes: frequency*

7.5.3. *Convergence of the NMV estimator*

It is of interest to study the asymptotic properties of the NMV estimator, which justify its utilization. Let us consider the intermediate expression of the estimator given by equation [7.38] and which can be written in the following form:

$$\mathbf{R}_x \mathbf{A}_{f_c} = \left(\frac{1}{T_s} S_{NMV}\left(f_c\right) \right) \mathbf{A}_{f_c} \qquad [7.45]$$

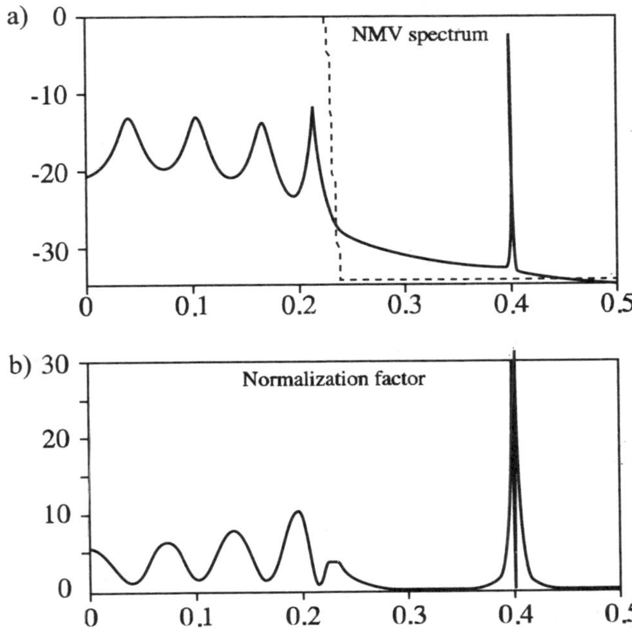

Figure 7.11. *Spectral analysis by the NMV estimator of the same signal as that used for Figure 7.5. a) Superposition of the true spectrum (in thin line) and of the NMV estimation at order 12. b) Normalization factor in Hz. Vertical axes: spectra in dB. Horizontal axes: frequency in Hz*

This equation is an eigenvalue equation. At each frequency f_c, the impulse response calculated by the Capon estimator is an eigenvector of the signal covariance matrix. The associated eigenvalue is the value of the spectrum estimated by the normalized NMV estimator at this frequency f_c. This link with the eigenvalue theory [LAG 86] indicates that the NMV method is essentially a pure frequency estimation method because its calculation comes back to a calculation of eigenvalues. Figure 7.11 illustrates this point of view.

This equation has a second point of interest. It makes it possible to study the convergence of the NMV estimator. Let us consider the following theorem [GRE 58].

Let $\{\lambda_0, \lambda_1, \ldots, \lambda_{M-1}\}$ be the set of eigenvalues and $\{v_0, v_1, \ldots, v_{M-1}\}$ the set of eigenvectors of the autocorrelation matrix $\mathbf{R}_x$, this matrix being Toeplitz matrix, if $M \rightarrow \infty$, then, for $I = 0, \ldots, M-1$, we have the following convergence property:

$$\lambda_1 \rightarrow S_x\left(i / MT_s\right)$$

$$\mathbf{v}_1 \rightarrow \frac{1}{\sqrt{M}}\left(1, e^{2\pi j / MT_s}, \ldots, e^{2\pi j(M-1)i / MT_s}\right) \qquad [7.46]$$

Thus, we deduce the convergence property of the NMV estimator when considering the eivenvalue equation [7.45] and applying the theorem [7.46].

If $M \rightarrow \infty$, then for any frequency f_c multiple of $1/MT_s$:

$$\lambda_{f_c} = \frac{1}{T_s} S_{NMV}(f_c) \rightarrow S_x(f_c)$$

$$\mathbf{v}_{f_c} = \mathbf{A}_{f_c} \qquad \rightarrow \frac{1}{\sqrt{M}} \mathbf{E}_{f_c}$$

This theorem is significant because it indicates that when the filter order M tends to the infinite, the NMV estimator converges towards the true value of the spectrum for all the frequencies multiple of $1/MT_s$.

As a conclusion, at a finite order M, the NMV estimator is adapted to the estimation of pure frequencies without giving a correct estimation in amplitude. It brings a spectral refinement in relation to the MV estimator, which makes it possible for it to separate close frequencies. In order to improve the estimation, it is necessary to increase the order because, at the infinite, the NMV estimator converges towards the true spectrum.

7.5.4. *Generalized MV estimator*

In order to improve the convergence of the NMV estimator, M. A. Lagunas [LAG 86] proposed a generalization of this estimator that we will note by $S_{GMV}(f)$ and which is written (q being a positive integer):

$$S_{GMV}(fc) = T_s \frac{\mathbf{E}_{f_c}^H \mathbf{R}_x^{-q+1} \mathbf{E}_{f_c}}{\mathbf{E}_{f_c}^H \mathbf{R}_x^{-q} \mathbf{E}_{f_c}} \qquad [7.47]$$

The justification of this generalized estimator comes from the fact that, when a positive defined matrix is raised to a power higher than 1, the contrast between the eigenvalues is increased. This estimator is also defined as a data filtering [MAR 95].

The impulse response, noted by $\mathbf{A}_{f_{c,q}}$, does not apply directly on the signal to be analyzed but on the preprocessed signal so that the signal correlation matrix should be raised to the power q at the end of the preprocessing. Instead of equation [7.4], the filter output power that we will note by P_q is written:

$$P_q = \mathbf{A}_{f_{c,q}}^H \mathbf{R}_x^q \mathbf{A}_{f_{c,q}} \tag{7.48}$$

The minimization of this equation with the same constraint as the MV estimator (equation [7.1]) leads to the following expression of $\mathbf{A}_{f_{c,q}}$:

$$\mathbf{A}_{f_{c,q}} = \frac{\mathbf{R}_x^{-q}\mathbf{E}_{f_c}}{\mathbf{E}_{f_c}^H \mathbf{R}_x^{-q}\mathbf{E}_{f_c}} \tag{7.49}$$

This estimator is a generalization of the MV estimators. Actually, if the filter is designed this way, it is directly applied on the signal, the output power is the generalized estimator noted $\boldsymbol{P}_{q,G}$ (f_c) and is defined by:

$$P_{q,G}(f_c) = \mathbf{A}_{f_{c,q}}^H \mathbf{R}_z \mathbf{A}_{f_{c,q}} \tag{7.50}$$

According to the value of q, we find the different estimators:

- $q = 0$ $\quad P_{0,G}(f_c) = \dfrac{\mathbf{E}_{f_c}\mathbf{R}_x\mathbf{E}_{f_c}}{M^2}$ given the periodogram estimator;
- $q = 1$ $\quad P_{1,G}(f_c) = P_{MV}(f_c)$;
- $q = 2$ $\quad P_{2,G}(f_c) = S_{NMV}(f_c)$;
- $q \geq 2$ $\quad P_{q,G}(f_c) = S_{GMV}(f_c)$.

For an increasing value of q and at a finite order M, the generalized estimator provides a correct value of the position of the peaks in frequency. Experimentally, the generalized estimator converges quite quickly and sometimes it is not necessary to use the orders q higher than 5. However, the higher q is, the more disturbed the amplitude of the peaks will be. This generalization is thus supported by the minimization techniques under constraints, which can include other estimators than those presented in this section [MAR 88, MAR 95].

It can be interesting to couple this frequency estimation with a power estimator, such as the Capon estimator. This leads to the hybrid estimators. The hybrid methods couple the Capon estimator with frequency estimators, either the NMV (LAGCAP estimator) or the autoregressive estimator (ARCAP). These estimators are presented in [DON 04] as well as in [LEP 98, PAD 96].

7.6. The CAPNORM estimator

The strong idea of the normalized minimum variance estimator presented in section 7.5 is to define a spectral density estimator starting from the power estimator given by the minimum variance. This idea is based on equation [7.35]. However, this equation makes the implicit hypothesis of the MV narrow band filters, a hypothesis that is not verified. It is of interest to propose another writing of this equation.

A solution consists of adapting the integration support to the filter shape. [DUR 00] proposes such an adaptation, which leads to a new estimator called CAPNORM.

The significant part of the frequency response is located around the filter frequency f_c. Let ΔB be the width of the lobe containing f_c, the contribution to the integral of equation [7.2] is close to 0 outside this band, either $S_x(f) \approx 0$, or $|Af_c(f)| \approx 0$. We can then write:

$$P_{MV}(f_c) \approx \int_{\Delta B} \left|A_{f_c}(f)\right|^2 S_x(f)\,df \qquad [7.51]$$

If we make the hypothesis that $S_x(f)$ is constant in the band ΔB, such as $S_x(f) \approx S_x(f_c)$, then equation [7.51] is written:

$$P_{MV}(f_c) \approx S_x(f_c) \int_{\Delta B} \left|A_{f_c}(f)\right|^2 df \qquad [7.52]$$

By substituting [7.6] into [7.52], we obtain the definition of a new estimator noted by $S_{CNorm}(f_c)$ [DUR 00]:

$$S_{CNorm}(f_c) = \frac{1}{\mathbf{E}_{fc}^H \mathbf{R}_x^{-1} \mathbf{E}_{fc} \beta_{CNorm}(f_c)} \qquad [7.53]$$

with $\beta_{CNorm}(f_c)$ defined by:

$$\beta_{CNorm}(f_c) \approx \int_{\Delta B} \left|A_{f_c}(f)\right|^2 df \qquad [7.54]$$

It remains to evaluate this local integral that we will numerically approximate by:

$$\beta_{CNorm}(f_c) \approx \Delta B \sup_{f \in \Delta B} \left|A_{f_c}(f)\right|^2 \qquad [7.55]$$

with the band $\Delta B = [f_-, f_+]$, which can be called an equivalent band of the filter at the frequency f_c, defined by:

$$\left|A_{f_c}(f_-)\right|^2 = \left|A_{f_c}(f_+)\right|^2 = \frac{1}{2}\left|A_{f_c}(f)\right|^2 \qquad [7.56]$$

The quantity $\beta_{CNorm}(f_c)$ should be considered as an integration support, which adapts to the shape of the frequency response around the frequency f_c of the filter, the band ΔB being the local equivalent band of the lobe containing this frequency.

Figure 7.12 illustrates the CAPNORM estimator on the same signal as that used for Figures 7.5 and 7.11. The peak of the exponential frequency is sharpened. The amplitude is always underestimated but with a weaker bias. The strong improvement is especially on the signal wide band. The estimation is of the true spectrum order, the oscillations being almost non-existent. The amplitude of the wideband spectrum is around the theoretical value, which is a considerable gain even in relation to the MV estimator (see Figure 7.13). The factor $\beta_{CNorm}(f_c)$ is weaker and more stable than the normalization factor $\beta_{NMV}(f_c)$. An experimental statistic study of this estimator is presented in [DUR 00].

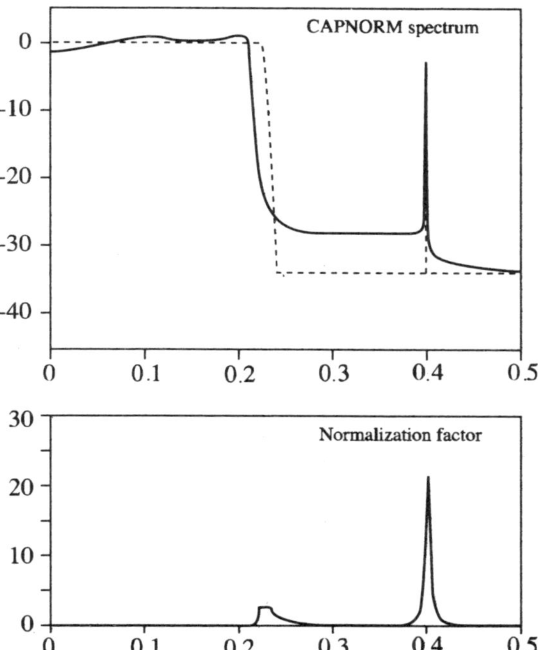

Figure 7.12. *Spectral analysis by the CAPNORM estimator of the same signal as that used for Figure 7.5. a) Superposition of the true spectrum (in thin line) and of the CAPNORM estimation at order 12. b) Normalization factor in Hz. Vertical axis: spectra in dB. Horizontal axes: frequency in Hz*

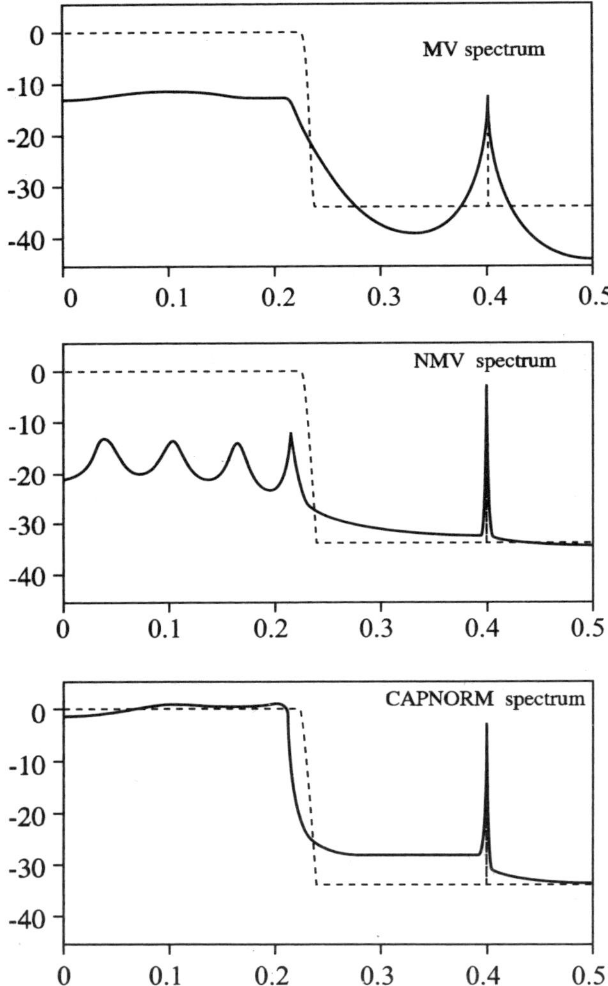

Figure 7.13. *Comparison of the MV, NMV and CAPNORM estimators, all at order 12, and the true spectrum (in thin line) for the spectral analysis of the same signal as that used for Figure 7.5. Horizontal axes: frequency in Hz. Vertical axes: spectra in dB*

7.7. Bibliography

[BAS 92] BASSEVILLE M., FLANDRIN P., MARTIN N., "Signaux non-stationnaires. Analyse temps-fréquence et segmentation", *Traitement du Signal,* vol. 9, no. 1, 1992, Supplement.

[BON 90] BONDANT L., MARS J., MARTIN N., *Analyse spectrale de signaux non-stationnaires par des méthodes temps-frequence: Fourier, Wigner-Ville lissée, AR et Lagunas evolutifs, méthodes hybrides,* SFA, Lyon, 1990.

[CAP 67] CAPON J., GREENFIELD, KOLKER R. J., "Multidimensional maximum-likelihood processing of a large aperture seismic array", *Proceedings of the IEEE,* vol. 55, p. 192-211, February 1967.

[CAP 69] CAPON J., "High-resolution frequency-wavenumber spectrum analysis", *Proceedings of the IEEE,* vol. 57, p. 1408-1418, August 1969.

[CAP 70] CAPON J., GOODMAN N. R., "Probability distributions for estimates of the frequency-wavenumber spectrum", *Proceedings of the IEEE*, vol. 58, p. 1785-1786, 1970.

[CAP 71] CAPON J., "Correction to 'Probability distributions for estimates of the frequency-wavenumber spectrum'", *Proceedings of the IEEE*, vol. 59, p. 112, January 1971.

[DON 04] DONCARLI C., MARTIN N., *Décision dans le plan temps-fréquence,* IC2 series, Hermes Sciences/Lavoisier, Paris, 2004.

[DUR 99] DURNERIN M., Une stratégie pour l'interprétation en analyse spectrale. Détection et caractérisation des composantes d'un spectre, PhD Thesis, INPG, September 1999.

[DUR 00] DURNERIN M., MARTIN N., "Minimum variance filters and mixed spectrum estimation", *Signal Processing,* vol. 80, no. 12, p. 2597-2608, December 2000.

[FER 86] FERNANDEZ J., MARTIN N., "Adaptive spectral estimation by ML filtering", *Proc. of EUSIPCO 86,* p. 323-326, 1986.

[GOL 85] GOLUB G. H., LOAN C. F. V., *Matrix Computations,* Johns Hopkins University Press, US, Maryland, 1985.

[GRE 58] GRENANDER U., SZEGO G., *Toeplits forms and their applications,* University of California Press, Berkeley, US, 1958.

[GUE 85] GUEGUEN C., "Analyse de la parole par les méthodes de modélisation paramétrique", *Annales des Télécommunications,* vol. 40, no. 5-6, p. 253-269, 1985.

[HON 98a] HONGBIN L., STOICA P., JIAN L., "Capon estimation of covariance sequence", *Circuits Systems and Signal-Processing,* vol. 17, no. 1, p. 29-49, 1998.

[HON 98b] HONGBIN L., STOICA P., JIAN L., JAKOBSSON A., "On the performance analysis of forward-only and forward-backward matched filterbank spectral estimators", *Conference Record of the Thirty-First Asilomar Conference on Signals, Systems and Computers,* Los Alamitos, CA, US, IEEE Comput. Soc, 1998.

[IOA 94] IOANNIDIS E. E., "On the behaviour of a Capon-type spectral density estimator", *The Annals of Statistics,* vol. 22, no. 4, p. 2089-2114, 1994.

[JAK 00] JAKOBSSON A., STOICA P., "Combining Capon and APES for estimation of spectral lines", *Circuits, Systems, and Signal Processing,* vol. 19, no. 2, p. 159-169, 2000.

[JAN 99] JANSSON M., STOICA, "Forward-only and forward-backward sample covariances – a comparative study", *Signal Processing,* vol. 77, no. 3, p. 235-245, September 1999.

[KAY 88] KAY S. M., *Modern Spectral Estimation: Theory and Applications,* Prentice-Hall, Englewood Cliffs (NJ), US, 1988.

[LAC 71] LACOSS R. T., "Data adaptive spectral analysis methods", *Geophysics,* vol. 36, no. 4, p. 661-675, August 1971.

[LAC 77] LACOSS R. T., "*Aspects of Signal Processing*", Autoregressive and maximum likelihood spectral analysis methods, p. 591-615, G. Tacconi Eds, part. 2th edition, 1977.

[LAG 84] LAGUNAS M. A., GASULL A., "An improved maximum likelihood method for power spectral density estimation", *IEEE Trans. Acoust. Speech Signal Process.,* vol. 32, p. 170-172, February 1984.

[LAG 86] LAGUNAS M. A., SANTAMARIA M. E., GASULL A., MORENO A., "Maximum likelihood filters in spectral estimation problems", *Signal Processing,* vol. 10, no. 1, p. 19-34, January 1986.

[LAT 87] LATOMBE C., "Analyse spectrale évolutive – comparaison de plusieurs algorithmes sur l'ensemble de plusieurs signaux tests du GRECO SARTA", *GRETSI,* June 1987.

[LEP 98] LEPRETTRE B., MARTIN N., GLANGEAUD R., NAVARRE J. P., "Three-component signal characterization in the time, time-frequency, and polarization domains using the arcap method. Application to seismic detection of avalanches", *IEEE Transactions on Signal Processing,* vol. 46, no. 1, January 1998.

[LEV 64] LEVIN, Maximum likelihood array processing, semi annual tech. summary report on seismic discrimination, Mass. Inst. Tech. Lincoln Lab., Lexington, Mass., December 1964.

[LI 98] LI H., LI J., STOICA P., "Performance analysis of forward-backward matched-filterbank spectral estimators", *IEEE Transactions on Signal Processing,* vol. 46, no. 7, p. 1954-1966, July 1998.

[LIU 98a] LIU X., SHERMAN P. J., "Asymptotic statistical properties of the Capon MV spectral estimator for mixed spectrum processes", *Ninth IEEE Signal Processing Workshop on Statistical Signal and Array Processing,* New York, US, p. 328-331, 1998.

[LIU 98b] LIU Z. S., LI H., LI J., "Efficient implementation of Capon and APES for spectral estimation", *IEEE Transactions on Aerospace and Electronic Systems*, vol. 34, no. 4, p. 1314-1319, October 1998.

[MAR 88] MARTIN N., LACOUME J. L., SANTAMARIA M. E., "Unified Spectral Analysis", *EUSIPCO,* p. 179-182, 1988.

[MAR 95] MARTIN N., MARS J., MARTIN J., CHORIER C., "A Capon's time-octave representation. Application in room acoustics", *IEEE Trans. in Signal Processing,* vol. 43, no. 8, p. 1842-1854, August 1995.

[PAD 95] PADOVESE L., TERRIEZ J. M., MARTIN N., "Etude des phénomènes dynamiques dans les pylônes compression des téléphériques monocâble", *Mécanique Industrielle et Matériaux*, vol. 48, no. 4, December 1995.

[PAD 96] PADOVESE L., MARTIN N., TERRIEZ J. M., "Méthode temps-fréquence hybride arcap pour l'identification des caractéristiques dynamiques d'un pylône compression d'un téléphérique monocâble", *Traitement du Signal,* vol. 13, no. 3, September 1996.

[SHE 91] SHERMAN P. J., "On the family of ML spectral estimates for mixed spectrum identification", *IEEE Transactions on Signal Processing*, vol. 39, no. 3, March 1991.

[STO 98] STOICA P., JAKOBSSON A., LI J., "Matched-filter-bank interpretation of some spectral estimators", *Signal Processing,* vol. 66, no. 1, p. 45-59, April 1998.

[STO 99] STOICA P., HONGBIN L., JIAN L., "A new derivation of the APES filter", *IEEE Signal Processing Letters,* vol. 6, p. 205-206, August 1999.

Chapter 8

Subspace-based Estimators

8.1. Model, concept of subspace, definition of high resolution

8.1.1. *Model of signals*

Let $x_a(t)$ be an analogic signal made up of P complex sine waves in an additive noise:

$$x_a(t) = \sum_{i=1}^{P} \alpha_i e^{j2\pi f_i t} + b_a(t) \qquad [8.1]$$

where the α_i are centered complex random and independent variables of variances $\sigma_{\alpha i}^2$, where $b_a(t)$ is a centered, complex, white noise of variance $E = (|b_a(t)|^2) = \sigma^2$ independent from α_i. We suppose the noise and the signals to be circular so that the methods presented here will be based only on the first correlation functions of the different signals. Moreover, we will further work on the sampled signals $x(k) = x_a(kT_e)$ and $b(k) = b_a(kT_e)$ where T_e is a sampling period verifying the Shannon theorem for the complex sine waves.

Using the non-correlation hypotheses of the sine waves and of the noise, the autocorrelation function of the observations is written:

$$\gamma_{xx}(k,m,m') = \sum_{i=1}^{P}\sum_{l=1}^{P} E\left(\alpha_i \alpha_l^*\right) e^{j2\pi(f_i - f_l)kT_e} e^{j2\pi(mf_i - m'f_l)T_e} + \sigma^2 \delta_{m-m'} \qquad [8.2]$$

Chapter written by Sylvie MARCOS.

where $\delta_{m-m'}$ is the Kronecker symbol. If the α_i are non-correlated (for example, if $\alpha_i = A_i e^{j\phi_i}$ with the determinist A_i and the random ϕ_i uniformly distributed on $[0, 2\pi]$ and independent two by two), then $x(k)$ is stationary and its correlation coefficients are:

$$\gamma_{xx}(m-m') = \sum_{i=1}^{p} \sigma_{\alpha i}^2 e^{j2\pi f_i (m-m')T_e} + \sigma^2 \delta_{m-m'} \qquad [8.3]$$

and the correlation matrix of order M Γ_{xx} is written:

$$\Gamma_{xx} = \begin{bmatrix} \gamma_{xx}(0) & \cdots & \gamma_{xx}(M-1) \\ & \vdots \ddots \vdots & \\ \gamma_{xx}(M-1)^* & \cdots & \gamma_{xx}(0) \end{bmatrix} \qquad [8.4]$$

(* indicates the conjugate of a complex number). We will consider this case in the following.

In practice, it is not Γ_{xx} that will be used but its estimate:

$$\hat{\Gamma}_{xx} = \begin{bmatrix} \hat{\gamma}_{xx}(0) & \cdots & \hat{\gamma}_{xx}(M-1) \\ & \vdots \ddots \vdots & \\ \hat{\gamma}_{xx}(M-1)^* & \cdots & \hat{\gamma}_{xx}(0) \end{bmatrix} \qquad [8.5]$$

where:

$$\hat{\gamma}_{xx}(m) = \frac{1}{N} \sum_{n=-(N-1)/2}^{(N-1)/2} x(n+m)x^*(n) \qquad [8.6]$$

and N is supposed to be odd. This estimation of Γ_{xx} supposes the ergodicity of the signal $x(k)$.

8.1.2. *Concept of subspaces*

The methods presented in what comes next are based on the decomposition of the observation space in two subspaces, the signal subspace and the noise subspace. Actually, by introducing the vector $\mathbf{x}(k)$ of the M observations $\{x(k), x(k+1), \ldots, x(k+M-1)\}$, we easily verify that:

$$x(k)=\begin{bmatrix} x(k) \\ x(k+1) \\ \vdots \\ x(k+M-1) \end{bmatrix}=\mathbf{A}\mathbf{s}(k)+\mathbf{b}(k) \qquad [8.7]$$

with:

$$\begin{aligned} &\mathbf{A}=\left[\mathbf{a}_1,\ldots,\mathbf{a}_P\right] \\ &\mathbf{s}(k)=\left[\alpha_1 e^{j2\pi f_1 kT_e},\ldots,\alpha_P e^{j2\pi f_P kT_e}\right]^T \\ &\mathbf{b}(k)=\left[b(k),\ldots,b(k+M-1)\right]^T \end{aligned} \qquad [8.8]$$

where:

$$\mathbf{a}_i=\mathbf{a}(f_i)=\left[1,e^{j2\pi f_i T_e},e^{2j2\pi f_i T_e},\ldots,e^{j(M-1)2\pi f_i T_e}\right]^T \qquad [8.9]$$

We call $\mathbf{a}(f_i)$ the complex sinusoid vector at frequency f_i, $\mathbf{A}$ the matrix of dimension (M, P) of the complex sinusoid vectors and $\mathbf{s}(k)$ contains the amplitudes of the sine waves (^{T} and ^{H} indicate the transpose and respectively the transconjugate of a vector or of a matrix).

According to the model [8.7], we see that in the noise absence, the vector of the observations $x(k)$ of the space $\mathbb{C}_M$ of the vectors of dimension M of complex numbers belongs to the subspace of dimension P, $\text{esp}\{\mathbf{A}\}$ defined by the complex sinusoid vectors $\mathbf{a}(f_i)$ supposed to be linearly independent. In the noise presence, it is no longer the case. However, the information that interests us concerning the frequencies of the sine waves remains limited to this subspace called the signal subspace (the complex sinusoid space). We call noise subspace the subspace complementary to $\text{esp}\{\mathbf{A}\}$ in $\mathbb{C}_M$.

The subspace methods are based on the two fundamental hypotheses that follow: $P < M$ and *the matrix* $\mathbf{A}$ *of dimension* (M, P) *is of full rank.* This implies that the vectors $\mathbf{a}(f_i)$ for $i = 1, \ldots, P$, are not linearly dependent. It is easy to see, $\mathbf{A}$ being a Vandermonde matrix (see equations [8.8] and [8.9]), that this property is verified as soon as the f_i are all different. This is not the case in a large number of antenna processing problems where the source vectors $\mathbf{a}_i$ can have a form which is very different from that given in [8.9] according to the antenna geometry, the gains of sensors, the wave fronts, etc. (see [MAR 98], Chapter 23).

The methods based on this decomposition of the observation space are often called eigendecomposition-based methods because the first of them effectively proceeded this way. However, the subspace noise and the subspace signal concepts exist independently of any decomposition in eigenelements. Actually, there are methods that do not use this procedure. It is the case of the "linear" methods from which the propagator method originates, this method will be presented later.

8.1.3. *Definition of high-resolution*

The methods based on the decomposition of the observation space in signal and noise subspaces have, in most cases, the high-resolution property. We say that a method has a high-resolution when its asymptotic resolution power is theoretically infinite, in the sense where two signals, however close they may be in frequency terms, can be separated (we call this resolute or detected) whatever the signal on noise ratio (SNR) may be, provided that the number of samples *N* used for the estimation of the correlation matrix [8.5] and [8.6] of the observations tends to the infinite and that the model [8.4] is verified.

In order to better illustrate this concept of high-resolution in relation to the classic Fourier analysis, we rewrite the spectral estimators of the periodogram and of the correlogram here, making the same sizes of observations *M* and *N*, which were introduced in the preceding paragraphs, appear. Given *N* observations of the signal $x(k)$ partitioned in *L* adjacent sequences of *M* successive observations of $x(k)$, the Bartlett periodogram is given by [MAR 87]:

$$\begin{aligned} \hat{S}_{xx}^{per}(f) &= \frac{1}{L}\sum_{l=0}^{L-1}\hat{S}_{xx}^{(l)}(f) \\ \hat{S}_{xx}^{(l)}(f) &= \frac{1}{M}\left|\sum_{m=0}^{M-1}x(lM+m)\exp\{-j2\pi fmT_e\}\right|^2 \end{aligned} \qquad [8.10]$$

In the same way, the correlogram is written ([MAR 87]; see also section 5.2):

$$\hat{S}_{xx}^{corr}(f) = \sum_{m=-(M-1)/2}^{(M-1)/2} \hat{\gamma}_{xx}(m)\exp\{-j2\pi fmTe\} \qquad [8.11]$$

where $\hat{\gamma}_{xx}(m)$ is given in [8.6]. The high-resolution property of the methods of subspace comes from the fact that at fixed *M*, it is enough to increase *N* indefinitely so that the method distinguishes two sine waves, however close to each other they may be. This is not the case for the periodogram and for the correlogram methods where not only N but also M should increase indefinitely in order to have this result.

8.1.4. *Link with spatial analysis or array processing*

It is important to know that the methods established on the concept of signal and noise subspaces have their origin in the antenna processing domain where it is about extracting information, particularly spatial information, on P sources emitting or reflecting waves which propagate in a given physical environment starting from information measured simultaneously on a network of M sensors. It is easy to make a parallel between the problem of detection and estimation of P frequencies contained in a time series and the problem of detection and estimation of the arrival directions of P waves coming from dispersed sources in a given space. This parallelism is exact for the case of an array of omnidirectional sensors, aligned and equally spaced with equal gains, called uniform linear antenna. Inversely, all the spectral analysis methods could have been extended to the case of array processing [JOH 82].

In array processing, the vectors $\mathbf{a}_i$ are called source (directional) vectors and they take into account the propagation parameters of the waves in a considered environment, the localization parameters of the sources in relation to the antenna (therefore of the antenna form) and the reception parameters of the antenna sensors.

8.2. MUSIC

The MUSIC method (MUltiple SIgnal Characterization) was introduced at the same time in [SCH 79] and in [BIE 83] in antenna processing and it is a geometric approach of the problem. It can be seen as an extension of the Pisarenko method [PIS 73] in the case of the time series.

Let us take again the model [8.4] of the covariance matrix, which is defined, non-negative and hermitian, and let us consider its decomposition in eigenelements:

$$\boldsymbol{\Gamma}_{xx} = \sum_{m=1}^{M} \lambda_m \mathbf{v}_m \mathbf{v}_m^{\mathrm{H}} = \mathbf{V}\boldsymbol{\Lambda}\mathbf{V}^{\mathrm{H}} \qquad [8.12]$$

where $\boldsymbol{\Lambda} = diag\{\lambda_1,\ldots,\lambda_M\}$ and $\mathbf{V} = [\mathbf{v}_1,\ldots,\mathbf{v}_M]$. The eigenvalues λ_m are real and positive, arranged in descending order and the associate eigenvectors $\mathbf{v}_m$ are orthonormal. The following theorem determines the foundation of the MUSIC method.

THEOREM 8.1. *Let the following partition be*

$$\mathbf{V} = [\mathbf{V}_s, \mathbf{V}_b], \boldsymbol{\Lambda}_s = \text{diag}[\lambda_1, \ldots, \lambda_P], \boldsymbol{\Lambda}_b = \text{diag}[\lambda_{P+1}, \ldots, \lambda_M]$$

where $\mathbf{V}_s$ *and* $\mathbf{V}_b$ *are of respective dimensions* $M \times P$ *and* $M \times (M - P)$*, with* $P < M$*. Using the hypothesis that the covariance matrix* $\boldsymbol{\Gamma}_{ss}$ *of the amplitudes of the complex sine waves is of full rank P, the minimum eigenvalue of* $\boldsymbol{\Gamma}_{xx}$ *is* σ^2 *and is of multiplicity* $M - P$ *and the columns of* $\mathbf{V}_s$ *span the same subspace as* $\mathbf{A}$*, i.e.:*

$$\begin{aligned} \boldsymbol{\Lambda}_b &= \sigma^2 \mathbf{I} \\ \text{esp}\{\mathbf{V}_s\} &= \text{esp}\{\mathbf{A}\} \\ \text{esp}\{\mathbf{V}b\} &\perp \text{esp}\{\mathbf{A})\} \end{aligned} \qquad [8.13]$$

where $\perp$ *indicates the orthogonality.*

DEMONSTRATION. In the noise absence, the covariance matrix $\boldsymbol{\Gamma}_{xx}$ is written:

$$\boldsymbol{\Gamma}_{xx} = \mathbf{A}\boldsymbol{\Gamma}_{ss}\mathbf{A}^{\mathrm{H}} \qquad [8.14]$$

The matrices $\mathbf{A}$ and $\boldsymbol{\Gamma}_{ss}$ being of the respective dimensions (M, P) and (P, P), with $P < M$, the matrix $\boldsymbol{\Gamma}_{xx}$ of dimension (M, M), is of rank at most equal to P. In Theorem 8.1 we suppose that $\mathbf{A}$ and $\boldsymbol{\Gamma}_{xx}$ are of full rank, therefore $\boldsymbol{\Gamma}_{xx}$ is of rank P. On the other hand, $\boldsymbol{\Gamma}_{xx}$ covariance matrix, being a non-negative defined hermitian matrix, has eigenvalues which are real and positive or null. The hypothesis $P < M$ implies that $\boldsymbol{\Gamma}_{xx}$ has $M - P$ null eigenvalues and P strictly positive eigenvalues.

Let $\mathbf{v}$ be an eigenvector of $\boldsymbol{\Gamma}_{xx}$ associated with the null eigenvalue; then:

$$\boldsymbol{\Gamma}_{xx}\mathbf{v} = \mathbf{A}\boldsymbol{\Gamma}_{ss}\mathbf{A}^{\mathrm{H}}\mathbf{v} = 0 \qquad [8.15]$$

The matrices $\mathbf{A}$ and $\boldsymbol{\Gamma}_{xx}$ being of full rank, it comes:

$$\mathbf{A}^{\mathrm{H}}\mathbf{v} = 0$$

implying that $\mathbf{v}$ belongs to the kernel of the application associated to the matrix $\mathbf{A}^{\mathrm{H}}$, given esp $\{\mathbf{A}\}^{\perp}$, the subspace orthogonal to the subspace spanned by the columns of $\mathbf{A}$. If we call esp $\{\mathbf{V}_b\}$ the set of the eigenvectors associated with the null eigenvalue, then it immediately comes esp $\{\mathbf{V}_b\} \subset$ esp $\{\mathbf{A}\}^{\perp}$. Inversely, any vector of esp $\{\mathbf{A}\}^{\perp}$ is an eigenvector of $\mathbf{\Gamma}_{xx}$ associated with the null eigenvalue. Hence esp $\{\mathbf{V}_b\}$ = esp $\{\mathbf{A}\}^{\perp}$. This implies that the other eigenvectors of $\mathbf{\Gamma}_{xx}$ associated with strictly positive eigenvalues, given esp $\{\mathbf{V}_s\}$, are in esp $\{\mathbf{A}\}$. Then we have esp $\{\mathbf{V}_s\} \subset$ esp $\{\mathbf{A}\}$. Now, $\mathbf{\Gamma}_{xx}$ being hermitian, we know that there is an orthonormal basis of eigenvectors, which implies that the dimension of esp $\{\mathbf{V}_s\}$ is P. Hence esp $\{\mathbf{V}_s\}$ = esp $\{\mathbf{A}\}$.

In the presence of a white noise, the eigenvalues of:

$$\mathbf{\Gamma}_{xx} = \mathbf{A}\mathbf{\Gamma}_{ss}\mathbf{A}^{\mathrm{H}} + \sigma^2\mathbf{I}$$

are those of $\mathbf{A}\mathbf{\Gamma}_{ss}\mathbf{A}^{\mathrm{H}}$ increased by σ^2 and the eigenvectors are unchanged, hence the results of Theorem 8.1.

We indicate by signal subspace the subspace esp $\{\mathbf{V}_s\}$ spanned by the eigenvectors associated with the biggest P eigenvalues of $\mathbf{\Gamma}_{xx}$, and by noise subspace the subspace esp $\{\mathbf{V}_b\}$ spanned by the $M - P$ eigenvectors associated with the smallest eigenvalue σ^2. Thus, it immediately results that the complex sinusoid vectors $\mathbf{a}(f_i)$ are orthogonal to the noise subspace, given:

$$\mathbf{V}_b^{\mathrm{H}}\mathbf{a}(f_i) = 0, \ i = 1, \ldots, P \qquad [8.16]$$

Thus, the signal vectors corresponding to the searched frequencies are the vectors $\mathbf{a}(f)$ of the form given in [8.9] which are orthogonal to $\mathbf{V}_b$.

A number P of independent vectors of the form of [8.9] should exist in a subspace of dimension P of $\mathbb{C}_M$ so that the resolution of the non-linear system [8.16] gives the searched frequencies in a unique way. This is true provided that $P + 1 < M$.

THEOREM 8.2. *If $P + 1 < M$, any system of P vectors of the form* $\mathbf{a}(f)$ *given in [8.9], for different values of f, forms a free family of $\mathbb{C}_M$.*

DEMONSTRATION. In $\mathbb{C}_M$, an infinity of vectors of the form $\mathbf{a}(f)$ [8.9] exist when f scans the set of real numbers. However, any family of $I < M$ vectors $\{\mathbf{a}(f_1), \ldots, \mathbf{a}(f_I)\}$, where the parameters f_i are different, is free because it is easy to verify that the matrix formed by these vectors is of the Vandermonde type. Thus, in a subspace of dimension $I < M$, only I vectors of this type exist. Actually, if another one existed, noted $\{\mathbf{a}(f_{I+1})\}$ with $f_{I+1} \neq f_i$ for $i = 1, \ldots, I$ in this subspace of dimension I, it would be written as a linear combination of these I vectors and the system $\{\mathbf{a}(f_1), \ldots, \mathbf{a}(f_{I+1})\}$ would not be free, which is impossible if $I + 1 < M$. Therefore, in esp $\{\mathbf{A}\}$, subspace of dimension P, if $P + 1 < M$, only P vectors of the form $\mathbf{a}(f)$ exist, which are the complex sinusoid vectors $\mathbf{a}_i$, for $i = 1,\dots, P$, columns of $\mathbf{A}$.

In the case of array processing, because of the particular form of the source vectors $\mathbf{a}_i$, we generally cannot conclude on the uniqueness of the set of solutions [MAR 98, Chapter 23], and we make the hypothesis that only the source vectors are solutions of [8.16].

8.2.1. *Pseudo-spectral version of MUSIC*

The MUSIC algorithm is thus as follows. First, it is about determining the minimum eigenvalue of the covariance matrix [8.4] of the observations as well as the number of eigenvalues equal to this value. Next, we form the noise subspace of the eigenvectors associated with this minimum eigenvalue, and then we form the estimator:

$$S_{\text{MUSIC}}(f) = \frac{1}{\mathbf{a}^{\text{H}}(f)\mathbf{V}_b\mathbf{V}_b^{\text{H}}\mathbf{a}(f)} \qquad [8.17]$$

This estimator takes maximum values (theoretically infinite because its denominator measures the orthogonality between the noise subspace and the candidate vectors of the form [8.9]) for the true values of the frequencies present in the signal. It is often called pseudo-spectrum by analogy with the spectral estimators introduced in the preceding chapters, although it is not an energy spectrum.

In practice, the covariance matrix is replaced by its estimate $\hat{\mathbf{\Gamma}}_{xx}$ [8.5] so that the estimator [8.17] becomes:

$$S_{\text{MUSIC}}(f) = \frac{1}{\mathbf{a}^{\text{H}}(f)\hat{\mathbf{V}}_b\hat{\mathbf{V}}_b^{\text{H}}\mathbf{a}(f)} \qquad [8.18]$$

where $\hat{\mathbf{V}} = \left[\hat{\mathbf{V}}_s, \hat{\mathbf{V}}_b\right]$ is the partition of the eigenvectors of $\hat{\mathbf{\Gamma}}_{xx}$. Because $\hat{\mathbf{V}}_b$ differs from its true value $\mathbf{V}_b$, the estimator [8.18] will not make infinite peaks appear. However, the arguments of the maxima of [8.18] are an estimation of the searched frequencies. Actually, we could have been able to directly choose the denominator of [8.18] as estimator because it measures the orthogonality of a vector of analysis $\mathbf{a}(f)$ and of the noise subspace. It is for comparing this pseudo-spectrum to the other spectra that we have taken its inverse.

Let us note that the set of the eigenvectors of the covariance matrix [8.4] forming an orthonormed basis of $\mathbb{C}_M, \mathbf{V}_s\mathbf{V}_s^H$ and $\mathbf{V}_b\mathbf{V}_b^H$ represent the orthogonal projectors on the signal and noise subspaces respectively and they are such that:

$$\mathbf{V}_s\mathbf{V}_s^H = \mathbf{I} - \mathbf{V}_b\mathbf{V}_b^H \qquad [8.19]$$

It follows that maximizing [8.17] means maximizing:

$$\mathbf{a}^{\mathrm{H}}(f)\mathbf{V}_s\mathbf{V}_s^{\mathrm{H}}\mathbf{a}(f) \qquad [8.20]$$

which, in an equivalent manner, thus constitutes an estimator of the searched frequencies.

8.2.2. *Polynomial version of MUSIC*

Ideally, the P frequencies contained in the signal to be analyzed, $f_1, \ldots, f_P$ make to zero the following function:

$$\left(S_{\text{MUSIC}}(f)\right)^{-1} = \mathbf{a}^H(f)\mathbf{V}_b\mathbf{V}_b^H\mathbf{a}(f) \qquad [8.21]$$

which measures the orthogonality between an analysis vector $\mathbf{a}(f)$ and the noise subspace. By introducing the vector:

$$\mathbf{u}(z) = \left[1, z, z^2, \ldots, z^{M-1}\right]^T \qquad [8.22]$$

we have:

$$\mathbf{a}(f) = \mathbf{u}\left(e^{j2\pi f}\right) \qquad [8.23]$$

The searched frequencies are then the solution of $\left(S_{\text{MUSIC}}(f)\right)^{-1}(f_i) = 0$ which is rewritten:

$$\mathbf{u}^T(1/z_i)\mathbf{V}_b\mathbf{V}_b^H\mathbf{u}(z_i) = 0 \text{ for } z_i = e^{j2\pi f_i} \qquad [8.24]$$

In other words, the frequencies contained in the signal to be analyzed are theoretically the arguments of the P roots situated on the unit circle of:

$$P(z) = \mathbf{u}^T(1/z)\mathbf{V}_b\mathbf{V}_b^H\mathbf{u}(z) \qquad [8.25]$$

The uniqueness of the P roots on the unit circle for determining the P frequencies result from Theorem 8.2. It is important to note that $P(z)$ possesses $2M-2$ symmetric roots in relation to the unit circle, i.e. if z_i is the solution, then $1/z_i$ is also the solution. Thus, there are $M-1$ roots of module less than or equal to 1 (inside the unit circle) and the searched frequencies thus correspond to the P roots which are on the unit circle. In the practical case, we have only $\hat{\Gamma}_{xx}$ and we thus choose as estimate the P roots which are closest to the unit circle.

The relation between the polynomial version of MUSIC and the pseudo-spectral version is established by noting that:

$$P\left(e^{j2\pi f}\right)=a\prod_{i=1}^{M-1}\left|e^{j2\pi f}-z_i\right|^2 \qquad [8.26]$$

where a is a constant. $P(e^{j2\pi f})$ is thus the sum of the squares of the distances of $e^{j2\pi f}$ with the roots of $P(z)$; it is minimal when $e^{j2\pi f}$ is aligned with a root close to the unit circle. We show that the performance of the pseudo-spectral and polynomial versions of MUSIC is the same [MAR 98, Chapter 9]. The interest of this version resides in the complexity. Searching the zeros of a polynomial is much less costly in calculation than searching the peaks in a pseudo-spectrum.

The MUSIC method is known for its superior performance, especially in resolution, compared to the other methods. In particular, contrary to the methods presented before, the estimates of the pure frequencies obtained by the MUSIC method converge to their true values while the number of samples N used in the estimation of the covariance [8.5] and [8.6] tends to infinity. The MUSIC method is thus a high-resolution method.

However, the limitations of the MUSIC method are important. First, the MUSIC method is based on a precise modeling of the noise by supposing that the covariance matrix of the noise is known with a multiplying factor. In the case of array processing, the MUSIC method is also less robust with a poor knowledge of the propagation model (i.e. the form **a** of the source vectors) than the beamforming method (equivalent to the periodogram method) or the minimum variance method. The MUSIC method also requires a determination of the number of complex sine waves before building the estimator of the frequencies. The computational complexity of the MUSIC method, particularly for the search of the eigenelements of the covariance matrix, of the order of M^3 is an important constraint, which pushes the searchers to conceive less complex algorithms (see the section on linear methods). Finally, we will refer to [MAR 98, Chapters 9 and 11] for the study of the performance and of the robustness of the MUSIC method.

8.3. Determination criteria of the number of complex sine waves

The theory tells us that the number of complex sine waves, given the order of the model, is equal to the number of eigenvalues strictly higher than the smallest eigenvalue of the covariance matrix of the observed signals. However, because we have only a finite number of samples of the signal to analyze and thus only an estimate of the covariance and also because the computer programs of eigenelement research introduce numeric errors, the order of the model is difficult to determine. It seems delicate to choose a threshold without risking not to detecting a frequency contained in the signal. Actually, the performance of the MUSIC method is conditioned by the determination of the number of estimated complex sine waves.

In order to remedy this limitation, criteria resulted from the information theory were proposed in order to determine the order of a model [RIS 78, SCH 78]. We simply give here the two most well known criteria (Akaike, MDL). They are based on the likelihood function of the observations[1].

AIC criterion (Akaike Information Criterion). Akaike [AKA 73] proposed this criterion to determine the order of an AR model. It consists in minimizing the following quantity in relation to the supposed number p of complex sine waves:

$$\mathrm{AIC}(p) = -N\log\frac{\prod_{i=p+1}^{M}\hat{\lambda}_i}{\left(\frac{1}{M-p}\sum_{i=p+1}^{M}\hat{\lambda}_i\right)^{M-p}} + p(2M-p) \qquad [8.27]$$

where p is an estimation of P, N the number of observations and $\hat{\lambda}_i$ the eigenvalues of $\hat{\Gamma}_{xx}$ arranged in descending order. For this criterion, the estimation of the noise variance is given by the average of the $M-p$ smallest eigenvalues of the covariance matrix, given:

$$\hat{\sigma}^2(p) = \frac{1}{M-p}\sum_{i=p+1}^{M}\hat{\lambda}_i$$

MDL criterion (Minimum Description Length). This criterion is inspired from [RIS 78, SCH 78] to determine the order of a model. The MDL criterion differs from the AIC criterion by the second term, which is introduced:

$$\mathrm{MDL}(p) = -N\log\frac{\prod_{i=p+1}^{M}\hat{\lambda}_i}{\left(\frac{1}{M-p}\sum_{i=p+1}^{M}\hat{\lambda}_i\right)^{M-p}} + \frac{1}{2}p(2M-p)\log N \qquad [8.28]$$

---

1 For more details see Chapter 6.

The MDL criterion is a consistent criterion, i.e. it converges to the true value of P when N tends to infinity. Comparisons of this criterion exist in [MAR 98, Chapter 6] and in [ZHA 89], for example. It is known that the AIC criterion tends to overestimate the number of complex sine waves, even at strong SNR, while the MDL criterion tends to underestimate this number at weak or average SNR and at finite number of observations N.

Numerous improvements of these criteria were proposed [MAR 98, Chapter 6]. We can mention the works in the antenna-processing domain [CHO 93], which make a joint estimation of the number of sources and of their direction of arrival. There are also works which are concerned with determining the number of sources when the noise is colored [LEC 89, ZHA 89].

However, let us note that, philosophically speaking, the problem of determining a threshold in order to discriminate a sine wave from the noise is delicate. Indeed, if we consider that the noise term includes everything that is not a useful signal, the noise can then be made up of many weak-power sine waves. That is why even if, theoretically, the MUSIC method separates what is white noise from what is pure sine wave, in practice, it will always be difficult to decide on the exact number of sine waves. In other words, a weak-power sine wave (compared to the others) risks not being detected.

8.4. The *MinNorm* method

This method is a variant of the MUSIC method. Indeed, the estimator [8.17] tests the orthogonality of an analysis vector supposed to be a complex sinusoid vector, with respect to all the vectors of $\mathbf{V}_b$ basis of the noise subspace. The *MinNorm* method consists in using only one vector $\mathbf{d}$ of this noise subspace. Thus, let us consider the following spectral estimator:

$$S_{\mathrm{MN}}(f) = \frac{1}{\mathbf{a}^{\mathrm{H}}(f)\mathbf{d}\mathbf{d}^{\mathrm{H}}\mathbf{a}(f)} = \frac{1}{\left|\mathbf{a}^{\mathrm{H}}(f)\mathbf{d}\right|^{2}} \qquad [8.29]$$

and in a manner equivalent to MUSIC, the associated polynomial estimator:

$$P_{\mathrm{MN}}(z) = \sum_{m=1}^{M} d_m z^{m-1} \qquad [8.30]$$

where the d_m are the coefficients of vector **d** and with:

$$\left(S_{\mathrm{MN}}(f)\right)^{-1} = \left|P_{\mathrm{MN}}\left(e^{j2\pi f}\right)\right|^2 \tag{8.31}$$

It was shown in [KUM 83] that, by choosing the vector **d** of minimum norm, with 1 as first component, the zeros of the estimating polynomial of P complex sine waves in noise associated with the noise were reduced inside the unit circle.

Searching the vector **d** of minimum norm which belongs to esp ($\mathbf{V}_b$) means searching the vector **w** so that:

$$\begin{aligned} &\mathbf{d} = \mathbf{V}_b \mathbf{w} \\ &J(\mathbf{w}) = \|\mathbf{d}\|^2 + \lambda\left(\mathbf{d}^{\mathrm{H}} e_1 - 1\right) \text{minimum} \end{aligned} \tag{8.32}$$

where $\mathbf{e}_1$ is a vector of first component equal to 1 with all the other components equal to zero. Minimizing:

$$J(\mathbf{w}) = \mathbf{w}^{\mathrm{H}} \mathbf{V}_b^{\mathrm{H}} \mathbf{V}_b \mathbf{w} + \lambda\left(\mathbf{w}^{\mathrm{H}} \mathbf{V}_b^{\mathrm{H}} \mathbf{e}_1 - 1\right) \tag{8.33}$$

with respect to **w**, by noting that the vectors of $\mathbf{V}_b$ being orthonormal, $\mathbf{V}_b^{\mathrm{H}} \mathbf{V}_b = \mathbf{I}$, leads to:

$$\mathbf{w} = -\frac{\lambda}{2} \mathbf{V}_b^{\mathrm{H}} \mathbf{e}_1 \tag{8.34}$$

The condition of first component equal to 1 gives:

$$\lambda = -\frac{2}{\mathbf{e}_1^{\mathrm{H}} \mathbf{V}_b \mathbf{V}_b^{\mathrm{H}} \mathbf{e}_1} \tag{8.35}$$

Finally, we obtain:

$$\mathbf{d} = \frac{\mathbf{V}_b \mathbf{V}_b^{\mathrm{H}} \mathbf{e}_1}{\mathbf{e}_1^{\mathrm{H}} \mathbf{V}_b \mathbf{V}_b^{\mathrm{H}} \mathbf{e}_1} \tag{8.36}$$

Thus, the frequencies included in the signal can be directly obtained by searching the zeros of the polynomial of degree $M-1$, $P_{\mathrm{MN}}(z)$. It is shown in [KUM 83] that the P zeros corresponding to the peaks in [8.29] are situated on the unit circle and that the $M-P$ other zeros remain limited inside the unit circle.

The MinNorm method is more sensitive than the MUSIC method to a weak number of observations for the estimation of the covariance matrix. We also show that MinNorm has a better separating power than MUSIC, but a larger estimation variance (see [MAR 98], Chapter 9).

8.5. "Linear" subspace methods

8.5.1. *The linear methods*

The methods called linear are an alternative to the MUSIC method, as far as they are based on the orthogonal subspaces but they do not require any eigendecomposition. They use only linear operations on the covariance matrix and have a smaller computational complexity than the MUSIC method. Moreover, they can be made adaptive for applications in real time or so as to follow a non-stationary context.

Among these methods that were developed in array processing, we mainly distinguish the propagator method [MAR 90a, MAR 95, MUN 91], BEWE (Bearing Estimation Without EigenDEcomposition) [YEH 86], SWEDE (Sub-space method Without EigenDecomposition) [ERI 73], which were unified and compared in [MAR 97]. Only the propagator method is presented here because it has been shown that it possesses the best performance.

8.5.2. *The propagator method*

This method is based on the partition of the matrix of the complex sinusoid vectors [8.8] in accordance with:

$$\mathbf{A} = \begin{bmatrix} \mathbf{A}_1 \\ \mathbf{A2} \end{bmatrix} \qquad [8.37]$$

where $\mathbf{A}_1$ and $\mathbf{A}_2$ are the matrices of dimension $P \times P$ and $(M-P) \times P$ respectively.

DEFINITION OF THE PROPAGATOR. *Using the hypothesis that* $\mathbf{A}_1$ *is non-singular, the propagator* $\mathbf{P}$ *is the unique operator from* $\mathbb{C}_{M-P}$ *to* $\mathbb{C}_P$ *defined by the equivalent relations:*

$$\mathbf{P}^{\mathrm{H}} \mathbf{A}_1 = \mathbf{A}_2 \qquad [8.38]$$

or

$$\mathbf{Q}^{\mathrm{H}}\mathbf{A} = \mathbf{0} \text{ with } \mathbf{Q} = \begin{bmatrix} \mathbf{P} \\ -\mathbf{I}_{M-P} \end{bmatrix} \qquad [8.39]$$

where $\mathbf{I}_{M\text{-}P}$ *and* $\mathbf{0}$ *are the identity matrix of dimension* $M - P$ *and the null matrix of dimension* $(M - P) \times P$ *respectively.*

In spectral analysis, $\mathbf{A}$ and $\mathbf{A}_1$ are Vandermonde matrices. This implies that $\mathbf{A}$ and $\mathbf{A}_1$ are of full rank as soon as the pure frequencies are different. In array processing, given the more general form of the matrix $\mathbf{A}$ and $\mathbf{A}_1$, this definition suggests points which will not be discussed further here (see [MAR 97]).

We easily notice that the relation [8.39] is similar to [8.16]. It means that the subspace spanned by the columns of the matrix $\mathbf{Q}$, $\text{esp}\{\mathbf{Q}\}$, is included in the noise subspace $\text{esp}\{\mathbf{V}_b\}$. Moreover, because $\mathbf{Q}$ contains the block $\mathbf{I}_{M\text{-}P}$, its $M - P$ columns are linearly independent. It results that:

$$\text{esp}\{\mathbf{Q}\} = \text{esp}\{\mathbf{V}_b\} = \text{esp}\{\mathbf{A}\}^{\perp} \qquad [8.40]$$

On the other hand, if we introduce the matrix of dimension (M, P):

$$\mathbf{Q}^{\perp} = \begin{bmatrix} \mathbf{I}_P \\ \mathbf{P}^{\mathrm{H}} \end{bmatrix} \qquad [8.41]$$

we have:

$$\text{esp}\{\mathbf{Q}^{\perp}\} = \text{esp}\{\mathbf{V}_s\} = \text{esp}\{\mathbf{A}\} \qquad [8.42]$$

Therefore, the propagator makes it possible to define the noise and signal subspaces using matrices $\mathbf{Q}$ and $\mathbf{Q}^{\perp}$.

An important difference in defining the noise and signal subspaces with the vectors of $\mathbf{V}_b$ or of $\mathbf{V}_s$ is that the vectors of $\mathbf{Q}$ and of $\mathbf{Q}^{\perp}$ are not orthonormal and that the matrices $\mathbf{Q}\mathbf{Q}^H$ and $\mathbf{Q}^{\perp}\mathbf{Q}^{\perp H}$ do not represent operators of projection of the observation space on the noise and signal subspaces, respectively, as it is the case for $\mathbf{V}_b\mathbf{V}_b^H$ and $\mathbf{V}_s\mathbf{V}_s^H$. It was shown in [MAR 97] that an orthonormalization of the matrices $\mathbf{Q}$ and $\mathbf{Q}^{\perp}$ in $\mathbf{Q}_O$ and $\mathbf{Q}_O^{\perp}$ respectively gives better performance of estimation.

When the noise subspace is determined using the propagator, the pseudospectrum:

$$S_{\text{propa}}(f) = \frac{1}{\mathbf{a}^H(f)\mathbf{Q}_O\mathbf{Q}_O^H\mathbf{a}(f)} \qquad [8.43]$$

yields an estimator of the frequencies. The solutions are the arguments of the maxima of [8.43].

8.5.2.1. *Propagator estimation using least squares technique*

The advantage of the propagator is the simple way in which we can obtain it starting from the covariance matrix.

In *the no noise case*, the covariance matrix of the data $\mathbf{\Gamma}_{xx}$ [8.4] becomes:

$$\mathbf{\Gamma}_{xx} = \mathbf{A}\mathbf{\Gamma}_{ss}\mathbf{A}^H \qquad [8.44]$$

By using the partition [8.37] of the matrix of the complex sinusoid vectors and by decomposing the covariance matrix:

$$\mathbf{\Gamma}_{xx} = \left[\mathbf{A}\mathbf{\Gamma}_{ss}\mathbf{A}_1^{\mathrm{H}}, \mathbf{A}\mathbf{\Gamma}_{ss}\mathbf{A}_2^{\mathrm{H}}\right] = [\mathbf{G}, \mathbf{H}] \qquad [8.45]$$

where **G** and **H** are the matrices of dimension $M \times P$ and $M \times (M-P)$ respectively, the definition [8.38] implies:

$$\mathbf{H} = \mathbf{GP} \qquad [8.46]$$

This important relation means that the last $M - P$ columns of the covariance matrix without the noise belong to the subspace spanned by the first P columns. The propagator **P** is thus the solution of the overdetermined linear system [8.46].

Using the hypothesis that the sine waves are not totally correlated, a hypothesis which is made in the model and which is also necessary for MUSIC, the matrix **G** is of full rank and the propagator is obtained by the least-squares solution:

$$\mathbf{P} = \left(\mathbf{G}^{\mathrm{H}}\mathbf{G}\right)^{-1}\mathbf{G}^{\mathrm{H}}\mathbf{H} \qquad [8.47]$$

In *the noisy case*, the decomposition [8.45] of the matrix [8.4] is always possible, but the relation [8.46] is no longer valid. An estimation $\hat{\mathbf{P}}$ of the propagator can be obtained by minimizing the following cost function:

$$J(\hat{\mathbf{P}}) = \left\|\mathbf{H} - \mathbf{G}\hat{\mathbf{P}}\right\|^2 \qquad [8.48]$$

where || . || is the Frobenius norm. The minimization of [8.48] leads to:

$$\hat{\mathbf{P}} = \left(\mathbf{G}^H\mathbf{G}\right)^{-1}\mathbf{G}^H\mathbf{H} \qquad [8.49]$$

Thus, obtaining the propagator and therefore determining the matrix **Q** which leads to the noise subspace and to the pseudo-spectrum [8.43], require the inversion of a matrix of dimension P, which is less complex in calculations than the eigendecomposition of a matrix of dimension M. The reduction of the calculations compared to the MUSIC method is in $O(P/M)$ [MAR 95]. The higher the number of observations, compared to the number of frequencies to estimate, the bigger this gain. Moreover, an adaptive version, capable of pursuing possible variations in time of the noise subspace, was given in [MAR 96], limiting again the calculation complexity. The diminution of the complexity is then in $O(P^2/M^2)$ compared to MUSIC. Let us finally note that this estimation of the propagator does not require that the noise is white. However, if the noise is white, we can find a joint estimation of the variance of the noise and of the propagator, as is indicated in the following paragraph.

8.5.2.2. *Determination of the propagator in the presence of a white noise*

Using the hypothesis that the noise is spatially and temporally white of variance σ^2, **P** can be extracted starting from $\mathbf{\Gamma}_{xx}$ by introducing the following partition of the modified covariance matrix:

$$\mathbf{\Gamma}_{xx}(\delta) = \mathbf{\Gamma}_{xx} - \delta\mathbf{I}_M = \left[\mathbf{G}(\delta), \mathbf{H}(\delta)\right] \qquad [8.50]$$

where δ is a positive scalar and where $\mathbf{G}(\delta)$ and $\mathbf{H}(\delta)$ are sub-matrices of the respective dimensions $M \times P$ and $M \times (M - P)$. It is shown in [MAR 90a] that $\left(\hat{\mathbf{P}} = \mathbf{P}, \delta = \sigma^2\right)$ is the unique solution of:

$$\mathbf{H}(\delta) = \mathbf{G}(\delta)\hat{\mathbf{P}} \qquad [8.51]$$

The following proposition was established [MAR 90a].

PROPOSITION. *Under the hypothesis that the matrix* $\mathbf{A}_2$, *of dimension* $(M — P) \times P$ *is of rank* P, $\left(\hat{\mathbf{P}} = \mathbf{P}, \delta = \sigma^2\right)$ *is the unique solution of (8.51) if and only if* $M - P > P$.

Let us note that the bigger $M - P$ compared to P, the less constraining this additional hypothesis on $\mathbf{A}_2$, i.e. when the number of observations is bigger than the number of frequencies to estimate.

The estimation of the noise variance is obtained using the following comments. By partitioning $\mathbf{\Gamma}_{xx}$ of [8.4] in accordance with:

$$\mathbf{\Gamma}_{xx} = \begin{bmatrix} \mathbf{G}_1 & \mathbf{H}_1 \\ \mathbf{G}_2 & \mathbf{H}_2 \end{bmatrix} \begin{matrix} \}P \\ \}M-P \end{matrix} \qquad [8.52]$$

where $\mathbf{G}_1$, $\mathbf{G}_2$, $\mathbf{H}_1$ and $\mathbf{H}_2$ are matrices of dimensions $P \times P$, $(M-P) \times P$, $P \times (M-P)$ and $(M-P) \times (M-P)$ respectively, we have:

$$\begin{cases} \mathbf{G}_1 = \mathbf{A}_1 \mathbf{\Gamma}_{ss} \mathbf{A}_1^H + \sigma^2 \mathbf{I}_P \\ \mathbf{G}_2 = \mathbf{A}_2 \mathbf{\Gamma}_{ss} \mathbf{A}_1^H \\ \mathbf{H}_1 = \mathbf{A}_1 \mathbf{\Gamma}_{ss} \mathbf{A}_2^H \\ \mathbf{H}_2 = \mathbf{A}_2 \mathbf{\Gamma}_{ss} \mathbf{A}_2^H + \sigma^2 \mathbf{I}_{M-P} \end{cases} \qquad [8.53]$$

It is then easy to see that:

$$\sigma^2 = \frac{\operatorname{tr}\{\mathbf{H}_2 \mathbf{\Pi}\}}{\operatorname{tr}\{\mathbf{\Pi}\}} \qquad [8.54]$$

where $\mathbf{\Pi} = \mathbf{I}_{M-P} - \mathbf{G}_2 \mathbf{G}_2^{\dagger} = \mathbf{I}_{M-P} - \mathbf{A}_2 \mathbf{A}_2^{\dagger}$ and where tr {.} indicates the trace operator of a matrix and $\dagger$ the pseudo inverse.

A possible estimation of σ^2 is thus the following [MAR 90b, MAR 97]:

$$\hat{\sigma}^2 = \text{Ré}\left\{ \frac{\operatorname{tr}\{\hat{\mathbf{H}}_2 \hat{\mathbf{\Pi}}\}}{\operatorname{tr}\{\hat{\mathbf{\Pi}}\}} \right\} \qquad [8.55]$$

where $\hat{\mathbf{\Pi}} = \mathbf{I} - \hat{\mathbf{G}}_2 \hat{\mathbf{G}}_2^{\dagger}$ and where $\hat{\mathbf{H}}_2, \hat{\mathbf{G}}_2$ enter the partition of $\hat{\mathbf{\Gamma}}_{xx}$:

$$\hat{\mathbf{\Gamma}}_{xx} = \begin{bmatrix} \hat{\mathbf{G}}_1 & \hat{\mathbf{H}}_1 \\ \hat{\mathbf{G}}_2 & \hat{\mathbf{H}}_2 \end{bmatrix} \qquad [8.56]$$

When the noise variance is estimated, it can then be substracted from the covariance matrix Γ_{xx} so that the propagator is then extracted from [8.51]:

$$\hat{\mathbf{P}} = \mathbf{G}\left(\hat{\sigma}^2\right)^{\dagger} \mathbf{H}\left(\hat{\sigma}^2\right) \qquad [8.57]$$

Let us note that this estimation of the noise variance was taken again in [STO 92] and the authors have especially shown that this method was less complex than the search of the smallest eigenvalue of $\mathbf{\Gamma}_{xx}$. Actually, we evaluate here at $(M - P)\ P^2 + P^3/3$ the number of multiplications necessary to the calculation of $\mathbf{\Pi}$, then at $(M-P)^2$ the number of multiplications necessary to the calculation of $\hat{\sigma}^2$.

Along the same line, a joint estimation of $(P, \sigma^2, \mathbf{P})$ was obtained in [MAR 90a]. It should be noted that the joint determination of the number of complex sine waves, of the noise variance and of the propagator (and therefore of the noise and signal subspaces) remains less complex in calculation than the search of the eigenvalues of a matrix of dimension $M \times M$ and this being all the more so as the number of observations is bigger compared to the number of frequencies to estimate. The asymptotic performance of the propagator method was analyzed and compared to those of MUSIC and of the other "linear" methods in [MAR 98, Chapter 10]. There it is shown that the propagator method has performance which is superior to the other "linear" methods and performance equivalent to MUSIC.

8.6. The ESPRIT method

The ESPRIT method (Estimation of Signal Parameters via Rotational Invariance Techniques) [ROY 86] initially designed for the array processing can be applied to spectral analysis. Its main advantage is that it considerably reduces the cost in calculations of the estimate of the frequencies. Actually, this technique, rather than requiring minima or maxima in a pseudospectrum, uses the search of the eigenvalues of a matrix of dimension $P \times P$ from where we can directly extract the estimates of the pure frequencies contained in the signal. In array processing, this complexity reduction is obtained with the constraint that the antenna possesses an invariant form by translation (*translational displacement invariance*). In other words, the antenna is made up of doublets of sensors obtained by the same translation. It is the case particularly when the antenna is uniform linear (ULA model). We will directly give the application of the ESPRIT method to the spectral analysis and for the more general case of the antenna processing we recommend the reader to refer to [MAR 98, Chapter 3].

Let us consider two vectors, each of $M-1$ observations:

$$\bar{\mathbf{x}}(k)=\begin{bmatrix} x(k) \\ x(k+1) \\ \vdots \\ x(k+M-2) \end{bmatrix} \quad \text{and} \quad \bar{\mathbf{x}}(k+1)=\begin{bmatrix} x(k+1) \\ x(k+2) \\ \vdots \\ x(k+M-1) \end{bmatrix} \qquad [8.58]$$

According to [8.7], it is easy to see that we have:

$$\begin{aligned} \bar{\mathbf{x}}(k) &= \bar{\mathbf{A}}\mathbf{s}(k)+\bar{\mathbf{b}}(k) \\ \bar{\mathbf{x}}(k+1) &= \bar{\mathbf{A}}\mathbf{s}(k+1)+\bar{\mathbf{b}}(k+1) \end{aligned} \qquad [8.59]$$

where:

$$\begin{aligned} \bar{\mathbf{A}} &= \left[\bar{\mathbf{a}}_1,\ldots,\bar{\mathbf{a}}_P\right] \\ \mathbf{s}(k) &= \left[\alpha_1 e^{j2\pi f_1 kT_e},\ldots,\alpha_P e^{j2\pi f_P kT_e}\right]^T \\ \bar{\mathbf{b}}(k) &= \left[b(k),\ldots,b(k+M-2)\right]^T \end{aligned} \qquad [8.60]$$

and:

$$\bar{\mathbf{a}}_i = \bar{\mathbf{a}}(f_i) = \left[1, e^{j2\pi f_i T_e}, e^{2j2\pi f_i T_e},\ldots,e^{j(M-2)2\pi f_i T_e}\right]^T \qquad [8.61]$$

We note that:

$$\mathbf{s}(k+1) = \mathbf{\Phi}\mathbf{s}(k) \qquad [8.62]$$

where:

$$\mathbf{\Phi} = \text{diag}\left\{e^{j2\pi f_1 T_e},\ldots,e^{j2\pi f_P T_e}\right\} \qquad [8.63]$$

is a unitary matrix of dimension $P \times P$. Thus, we see that matrix $\mathbf{\Phi}$, which is often associated with a rotation operator, contains all the information about the frequencies to estimate.

We then introduce:

$$\mathbf{z}(k)=\begin{bmatrix}\bar{\mathbf{x}}(k)\\ \bar{\mathbf{x}}(k+1)\end{bmatrix}=\tilde{\mathbf{A}}\,\mathbf{s}(k)+\mathbf{n}(k) \qquad [8.64]$$

where:

$$\tilde{\mathbf{A}}=\begin{bmatrix}\bar{\mathbf{A}}\\ \bar{\mathbf{A}}\boldsymbol{\Phi}\end{bmatrix} \quad \text{and} \quad \mathbf{n}(k)=\begin{bmatrix}\bar{\mathbf{b}}(k)\\ \bar{\mathbf{b}}(k+1)\end{bmatrix} \qquad [8.65]$$

It is thus the structure of the matrix $\tilde{\mathbf{A}}$ of dimension $2MP \times P$, which will be exploited to estimate $\boldsymbol{\Phi}$ without having to know $\bar{\mathbf{A}}$.

It is easy to see that the covariance matrix $\boldsymbol{\Gamma}_{zz}$ of all the observations contained in $\mathbf{z}(k)$ is of dimension $2M\times 2M$ and is written:

$$\boldsymbol{\Gamma}_{zz}=E\left[\mathbf{z}(k)\mathbf{z}(k)^{\mathrm{H}}\right]=\tilde{\mathbf{A}}\boldsymbol{\Gamma}_{ss}\tilde{\mathbf{A}}^{\mathrm{H}}+\sigma^2\mathbf{I} \qquad [8.66]$$

where $\boldsymbol{\Gamma}_{ss}$ is the covariance matrix of dimension $P \times P$ of the amplitudes of the complex sine waves and $\mathbf{I}$ is the identity matrix of dimension $2M \times 2M$. Thus, the structure of the covariance matrix is identical to that of the observation [8.4] and we can then apply the theorem of the eigendecomposition and the definition of the signal and noise subspaces. In particular, let $\mathbf{V}_s$ be the matrix of dimension $2M\times P$ of the eigenvectors of the covariance matrix $\boldsymbol{\Gamma}_{xx}$, associated with the eigenvalues which are strictly superior to the variance of the noise σ^2, it results that the columns of $\mathbf{V}_s$ and the columns of $\tilde{\mathbf{A}}$ define the same signal subspace. There is thus a unique operator $\mathbf{T}$ as:

$$\mathbf{V}_s=\tilde{\mathbf{A}}\mathbf{T} \qquad [8.67]$$

By decomposing $\mathbf{V}_s$ as:

$$\mathbf{V}_s=\begin{bmatrix}\mathbf{V}_x\\ \mathbf{V}_y\end{bmatrix}=\begin{bmatrix}\bar{\mathbf{A}}\mathbf{T}\\ \bar{\mathbf{A}}\boldsymbol{\Phi}\mathbf{T}\end{bmatrix} \qquad [8.68]$$

where the matrices $\mathbf{V}_x$ and $\mathbf{V}_y$ are of dimension $M\times P$, we can see that the subspaces esp $\mathbf{V}_x$ = esp $\mathbf{V}_y$ = esp $\bar{\mathbf{A}}$ are the same. Let $\boldsymbol{\Psi}$ be the unique matrix $P \times P$ of transition from the basis of the columns of $\mathbf{V}_x$ to that of the columns of $\mathbf{V}_y$, we have:

$$\mathbf{V}_y=\mathbf{V}_x\boldsymbol{\Psi} \qquad [8.69]$$

Because $\mathbf{V}_x$ is of dimension $M \times P$ and of full rank P, it then results:

$$\boldsymbol{\Psi} = \left(\mathbf{V}_x^{\mathrm{H}} \mathbf{V}_x\right)^{-1} \mathbf{V}_x^{\mathrm{H}} \mathbf{V}_y \qquad [8.70]$$

On the other hand, by using the relation [8.69] and the partition [8.68] it results:

$$\overline{\mathbf{A}}\mathbf{T}\boldsymbol{\Psi} = \overline{\mathbf{A}}\boldsymbol{\Phi}\mathbf{T} \Rightarrow \overline{\mathbf{A}}\mathbf{T}\boldsymbol{\Psi}\mathbf{T}^{-1} = \overline{\mathbf{A}}\boldsymbol{\Phi} \qquad [8.71]$$

Because we have supposed that $\overline{\mathbf{A}}$ was of full rank, it results:

$$\mathbf{T}\boldsymbol{\Psi}\mathbf{T}^{-1} = \boldsymbol{\Phi} \qquad [8.72]$$

Therefore, the eigenvalues of $\boldsymbol{\Psi}$ are equal to the diagonal elements of the matrix $\boldsymbol{\Phi}$, and the columns of $\mathbf{T}$ are the eigenvectors of $\boldsymbol{\Psi}$.

It is on this latter relation that the ESPRIT method is based on. After having decomposed the matrix $\boldsymbol{\Gamma}_{zz}$ in eigenelements, the relation [8.70] makes it possible to obtain the operator $\boldsymbol{\Psi}$ whose eigenvalues are the diagonal elements of $\boldsymbol{\Phi}$ where the frequencies can be extracted from. In practice, only an estimation of the matrix $\boldsymbol{\Gamma}_{zz}$ as well as of its eigenvectors is available and the estimation of $\boldsymbol{\Psi}$ is obtained by replacing the relation [8.70] by the minimization of $\left\|\hat{\mathbf{V}}_y - \hat{\mathbf{V}}_x \boldsymbol{\Psi}\right\|^2$. The solution is again given by [8.70] where $\mathbf{V}_x$ and $\mathbf{V}_y$ are replaced by their estimates.

This least-squares process [GOL 80] consists in supposing that the estimation of $\mathbf{V}_x$ is perfect and that only $\hat{\mathbf{V}}_v$ has errors. The least-squares process thus aims to minimize these errors. The total least-squares method [GOL 80] makes it possible to take errors on the estimation of $\mathbf{V}_x$ and of $\mathbf{V}_y$ into account at the same time. The solution of the total least-squares criterion is given using the matrix of dimension $M \times 2P$, $\mathbf{V}_{xy} = [\mathbf{V}_x, \mathbf{V}_y]$. We then search the eigendecomposition of the matrix of dimension $M \times M$:

$$\mathbf{V}_{xy}^{\mathrm{H}} \mathbf{V}_{xy} = \mathbf{V}\boldsymbol{\Lambda}\mathbf{V}^{\mathrm{H}} \qquad [8.73]$$

We then partition the matrix of dimension $M \times M$ of the eigenvectors $\mathbf{V}$ in accordance with:

$$\mathbf{V} = \begin{bmatrix} \mathbf{V}_{11} & \mathbf{V}_{12} \\ \mathbf{V}_{21} & \mathbf{V}_{22} \end{bmatrix} \qquad [8.74]$$

It finally results:

$$\Psi = -\mathbf{V}_{12}\mathbf{V}_{22}^{-1} \quad [8.75]$$

Finally, let us note that a method of array processing, also applicable to the spectral analysis, using at the same time the technique of the propagator in order to obtain the signal subspace and the ESPRIT technique for the search of the pure frequencies starting from the signal subspace, was successfully proposed in [MAR 94]. The joint usage of these two techniques makes it possible to reduce the complexity of the calculations.

8.7. Illustration of the subspace-based methods performance

Here it is not about making a compared study of the performance of various spectral analysis methods but simply about illustrating the behavior of the main subspaces-based methods with respect to the more classic methods on a simple example.

Indeed, it is known that the methods of the periodogram and of the correlogram are limited by the parameter M (size of the window on which we make the Fourier transform [8.10] and [8.11]). The minimum variance method or Capon method [CAP 69] is limited by the signal to noise ratio. It is actually proved that for an infinite SNR, the Capon method is identical to the MUSIC method [PIL 89]. Finally, it is demonstrated that the inverse of the Capon spectral estimator is an average of the inverses of the spectral estimators of linear prediction of various orders [BUR 72]. This explains that the Capon method is more smoothed than the linear prediction method. These results will be illustrated later[2].

The simulations concern a signal having two complex sine waves whose frequencies are respectively equal to $f_1 = 0.15F_e$ and to $f_2 = 0.18F_e$ where F_e is the sampling frequency. The observed signal is that of [8.1] and [8.7] where the noise is characterized by its signal to noise ratio (SNR). Only the techniques using the pseudospectra are here represented. Figures 8.1 to 8.4 represent the average of the pseudospectra on 100 Monte-Carlo simulations obtained for the periodogram methods, minimum variance or the Capon method, linear prediction, propagator method, MUSIC and *MinNorm* for different values of the parameters M, SNR, N. We notice the superiority of the subspace-based methods compared to the classic methods even if the linear prediction method offers here a satisfactory behavior apart from the oscillations. In Tables 8.1 to 8.4, the different results of resolution,

2 A more detailed comparison of the performance of the periodogram and Capon methods is presented in Chapter 7.

average and standard deviations of the estimates are regrouped. In these tables, the resolution at $\pm\Delta$ for $\Delta = 0.01$, 0.005, 0.002 indicates the number of times (on 100 simulations) where two pure frequencies have been detected at the same time, each in an interval of $\pm\Delta$ around the true values. The averages and standard deviations are calculated on each of the detected sine waves.

8.8. Adaptive research of subspaces

Two essential calculation steps are necessary for the implementation of the subspace-based methods. The first consists in searching the signal or noise subspaces. The second consists in estimating the frequencies of the signal from the signal subspace or from the noise subspace. That can consist, as we have seen above, of searching either the minima or the maxima of a function called pseudo-spectrum, or the roots of a polynomial, or the eigenvalues of a matrix built with the help of a basis of the signal subspace.

Despite the different variants, we should keep in mind that this second calculation stage remains costly. However, this stage is also necessary for the more classic methods of spectral analysis.

The first calculation step mentioned above is specific to the subspace methods.

The search of the eigenvalues of a matrix is complex from a computational point of view and it is the more so as the dimension of the space of the observations increases. Moreover, in a non-stationary environment, this research should be reproduced at each new observation. In practice, $\hat{\Gamma}_{xx}$ is updated in an adaptive way by introducing the time variable k and by adopting the recursive relation:

$$\hat{\Gamma}_{xx}(k) = \beta\hat{\Gamma}_{xx}(k-1) + \alpha\mathbf{x}(k)\mathbf{x}^{H}(k) \qquad [8.76]$$

where β and α are weighting coefficients (forgotten factors).

A large number of adaptive algorithms have been developed in published works in order to estimate the matrix whose columns span the signal subspace in a recursive way (i.e. by using the calculations of the preceding moment each time a new data is measured). Classification tests of these algorithms have been given in the following works [BIS 92, COM 90, SCH 89, YAN 95]. Actually, it is possible to regroup them in three large families.

First, we find the algorithms that are inspired from the classic techniques of eigenvalue decomposition/singular value decomposition (EVD/SVD) as the QR iteration, the Jacobi algorithm, and the Lanczos algorithm [COM 90, FUH 88]. In this family we also find the algorithms derived from the power method [GOL 89, COM 90, KAR 84, OWS 78].

In the second family, we include the algorithms whose principle is based on the works of Bunch *et al.* [BUN 78] that analyze the updating of a hermitian matrix which was modified by the addition of a matrix of rank one [DEG 92, KAR 86, SCH 89].

The last family regroups the algorithms that consider the EVD/SVD as an optimisation problem with or without constraints: the gradient method [KAR 84, OWS 78, YAN 95], the conjugated gradient method [YAN 89], and the Gauss-Newton method [BAN 92, RED 82].

We should add to this classification a fourth family of "linear" methods that were presented above (see the propagator method) as not requiring any eigenelement decomposition and which can be made adaptive with a sensitive reduction of the complexity [ERI 73, MAR 96, YEH 87].

Another classification of the adaptive algorithms according to their complexity measured in number of complex operations, given the set of two multiplications and of one addition, was proposed in [MAR 98, Chapter 14], to which the interested reader is invited to refer. We distinguish the algorithms of complexity $O(M^2)$ (of the order of M^2 operations) and those of complexity $O(MP^2)$ and $O(MP)$. These magnitude orders are to be compared to a complexity in $O(M^3)$ for the eigenelement decomposition. This classification was also made in [COM 90].

However, it is necessary to keep in mind that the complexity alone is not a criterion sufficient to choose an algorithm. We should also take into account the convergence capacity of the algorithm (precision and velocity), its capacity of pursuing a non-stationary environment, its stability regarding the numeric errors as well as the hypothesis necessary to establish this algorithm. A comparison of these algorithms was made in [MAR 98, Chapter 14].

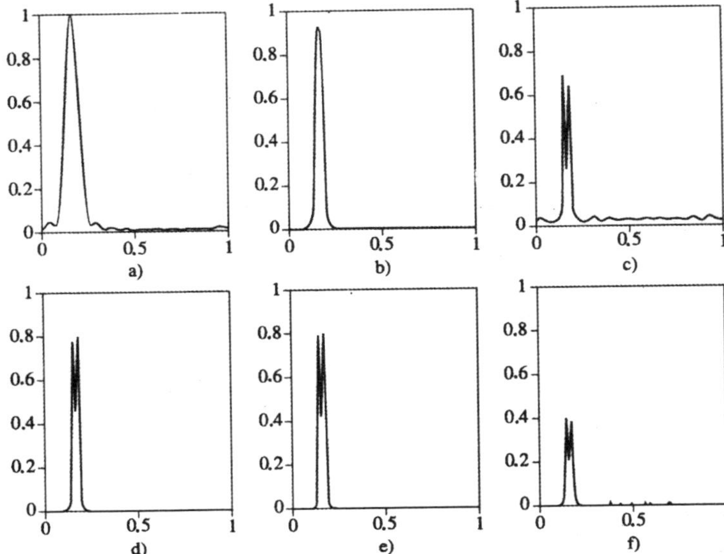

Figure 8.1. *Average of the pseudospectra over 100 Monte-Carlo simulations: a) Periodogram, b) Minimum variance method, c) Linear prediction, d) Propagator method, e) MUSIC, f) MinNorm.* $M = 12$, $N = 50$, $f_1/F_e = 0.15$, $f_2/F_e = 0.18$, $SNR = 10$ *dB*

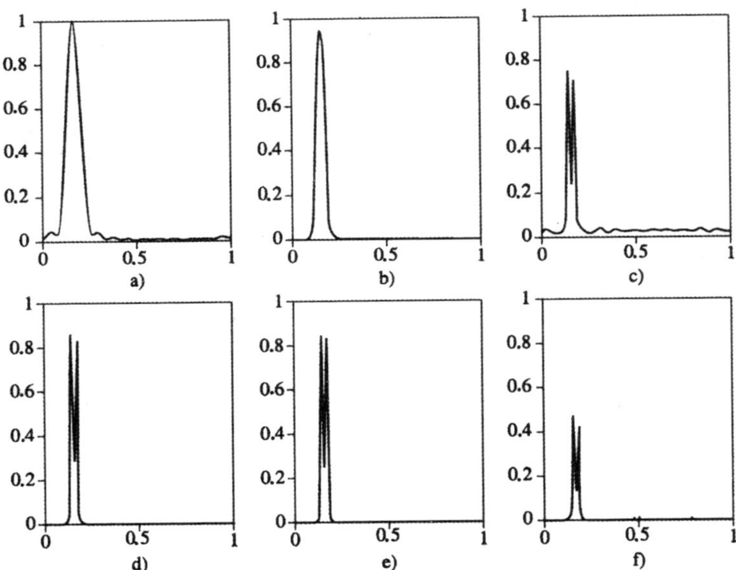

Figure 8.2. *Average of the pseudospectra over 100 Monte-Carlo simulations: a) Periodogram, b) Minimum variance method, c) Linear prediction, d) Propagator method, e) MUSIC, f) MinNorm.* $M = 12$, $N = 100$, $f_1/F_e = 0.15$, $f_2/F_e = 0.18$, $SNR = 10$ *dB*

Method	a)	b)	c)	d)	e)	f)
resolution at ±0.01	0	28	95	97	98	65
resolution at ±0.005	0	2	59	72	73	42
resolution at ±0.002	0	1	43	56	57	19
average 1	-	0.1548	0.1505	0.1514	0.1513	0.15
average 2	-	0.1739	0.1804	0.1787	0.1786	0.1795
standard deviation 1	-	0.1546	0.0964	0.0818	0.0812	0.1265
standard deviation 2	-	0.2095	0.0900	0.0831	0.0825	0.1118

Table 8.1. *Performance of the different methods: a) Periodogram, b) Minimum variance, c) Linear prediction, d) Propagator, e) MUSIC, f) MinNorm.* $M = 12$, $N = 50$, $f_1/F_e = 0.15$, $f_2/F_e = 0.18$, *SNR = 10 dB*

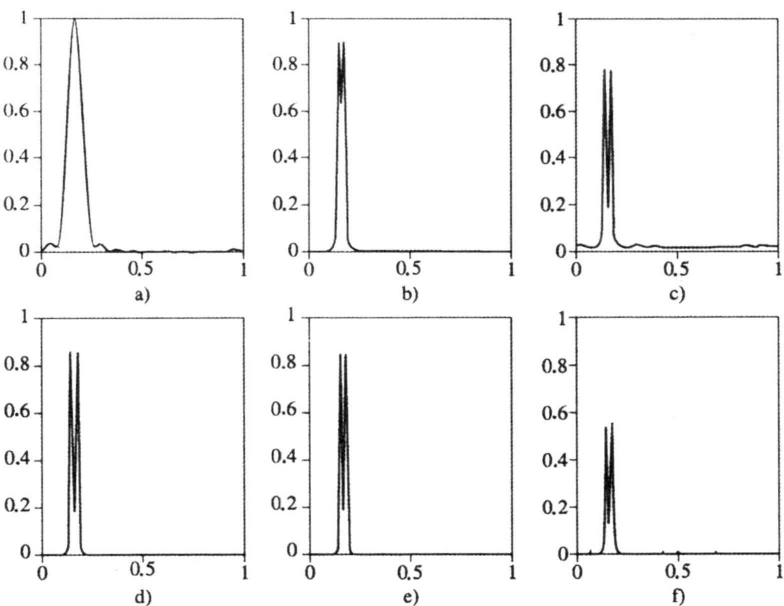

Figure 8.3. *Average of the pseudospectra over 100 Monte-Carlo simulations: a) Periodogram, b) Minimum variance method, c) Linear prediction, d) Propagator method, e) MUSIC, f) MinNorm.* $M = 12$, $N = 50$, $f_1/F_e = 0.15$, $f_2/F_e = 0.18$, *SNR = 15 dB*

Method	a)	b)	c)	d)	e)	f)
resolution at ±0.01	0	20	99	100	100	82
resolution at ±0.005	0	0	85	94	94	59
resolution at ±0.002	0	0	72	88	88	35
average 1	-	0.1550	0.1502	0.1503	0.1503	0.1498
average 2	-	0.1731	0.1799	0.1797	0.1797	0.1787
standard deviation 1	-	0.1581	0.0224	0.0412	0.0412	0.0943
standard deviation 2	-	0.2319	0.0574	0.0387	0.0387	0.1192

Table 8.2. *Performance of the different methods: a) Periodogram, b) Minimum variance, c) Linear prediction, d) Propagator, e) MUSIC, f) MinNorm.* $M = 12$, $N = 100$, $f_1/F_e = 0.15$, $f_2/F_e = 0.18$, *SNR = 10 dB*

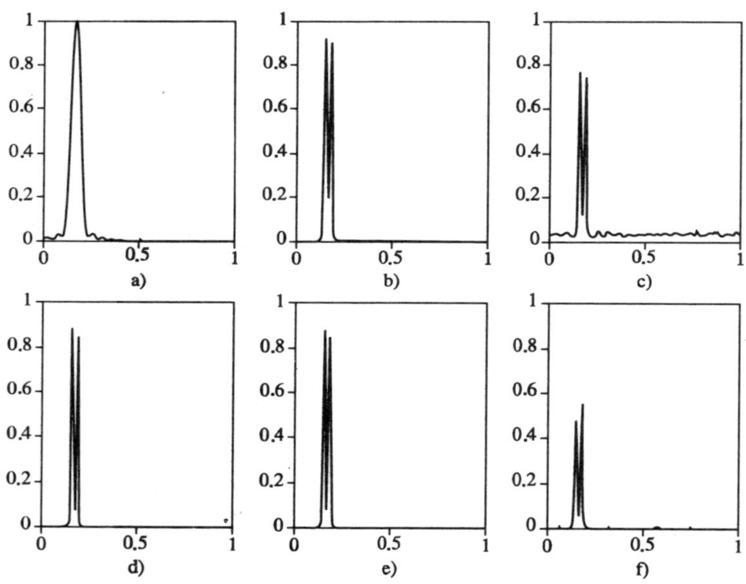

Figure 8.4. *Average of the pseudospectra over 100 Monte-Carlo simulations: a) Periodogram, b) Minimum variance method, c) Linear prediction, d) Propagator method, e) MUSIC, f) MinNorm.* $M = 12$, $N = 50$, $f_1/F_e = 0.15$, $f_2/F_e = 0.18$, *SNR = 10 dB*

Method	a)	b)	c)	d)	e)	f)
resolution at ±0.01	0	100	100	100	100	89
resolution at ±0.005	0	64	91	96	96	61
resolution at ±0.002	0	47	87	93	93	48
average 1	-	0.1518	0.1502	0.1502	0.1502	0.1501
average 2	-	0.1781	0.1801	0.1798	0.1798	0.1795
standard deviation 1	-	0.0949	0.0412	0.0316	0.0316	0.1034
standard deviation 2	-	0.0975	0.0424	0.0283	0.0283	0.1063

Table 8.3. *Performance of the different methods: a) Periodogram, b) Minimum variance, c) Linear prediction, d) Propagator, e) MUSIC, f) MinNorm.* $M = 12$, $N = 50$, $f_1/F_e = 0.15$, $f_2/F_e = 0.18$, $SNR = 15$ *dB*

Method	a)	b)	c)	d)	e)	f)
resolution at ±0.01	0	100	100	100	100	92
resolution at ±0.005	0	100	95	100	100	72
resolution at ±0.002	0	100	91	100	100	63
average 1	-	0.1500	0.1499	0.1500	0.1500	0.1500
average 2	-	0.1800	0.1801	0.1800	0.1800	0.1799
standard deviation 1	-	0.0000	0.0316	0.0000	0.0000	0.0728
standard deviation 2	-	0.0000	0.0360	0.0000	0.0000	0.0916

Table 8.4. *Performance of the different methods: a) Periodogram, b) Minimum variance, c) Linear prediction, d) Propagator, e) MUSIC, f) MinNorm.* $M = 12$, $N = 50$, $f_1/F_e = 0.15$, $f_2/F_e = 0.18$, $SNR = 10$ *dB*

8.9. Bibliography

[AKA 73] AKAIKE H., "Maximum likelihood identification of Gaussian autoregressive moving average models", *Biometrika,* vol. 60, p. 255-265, 1973.

[BAN 92] BANNOUR S., AZIMI-SADJADI M. R., "An adaptive approach for optimal data reduction using recursive least squares learning method", *Proc. IEEE ICASSP*, San Francisco, CA, vol. II, p. 297-300, 1992.

[BIE 83] BIENVENU G., KOPP L., "Optimality of high resolution array processing using the eigensystem approach", *IEEE Trans, on Acoustics, Speech and Signal Processing,* vol. ASSP-31, no. 10, p. 1235-1247, October 1983.

[BIS 92] BISCHOF C., SHROFF G., "On updating signal subspaces", *IEEE Trans. Signal Processing,* vol. 40, no. 10, p. 96-105, 1992.

[BUN 78] BUNCH J., NIELSEN C., SORENSON D., "Rank-one modification of the symmetric eigenproblem", *Numerische Math.,* vol. 31, p. 31-48, 1978.

[BUR 72] BURG J., "The relationship between maximum entropy spectra and maximum likelihood spectra", *Geophysics,* vol. 37, p. 375-376, 1972.

[CAP 69] CAPON J., "High-resolution frequency-wavenumber spectrum analysis", *Proceedings of the IEEE,* vol. 57, p. 1408-1418, 1969.

[CHO 93] CHOI J., SONG I., KIM S., KIM S., "A combined determination-estimation method for direction of arrival estimation", *Signal Processing,* vol. 30, p. 123-131, 1993.

[COM 90] COMON P., GOLUB G., "Tracking a few extreme singular values and vectors in signal processing", *Proceedings of the IEEE,* vol. 78, no. 8, p. 1327-1343, 1990.

[DEG 92] DEGROAT R. D., "Noniterative subspace tracking", *IEEE Trans. on Signal Processing,* vol. 40, p. 571-577, 1992.

[ERI 73] ERIKSSON A., STOICA P., SÖDERSTROM T., "On-line subspace algorithms for tracking moving sources", *IEEE Trans. Signal Processing,* vol. 42, no. 9, p. 2319-2330, 1973.

[FUH 88] FUHRMANN D., "An algorithm for subspace computation with applications in signal processing", *SIAM J. Matrix Anal. Appl,* vol. 9, no. 2, p. 213-220, 1988.

[GOL 80] GOLUB G., VAN LOAN C., "An analysis of the total least square problem", *SIAM Journal of Numerical Analysis*, vol. 17, no. 6, 1980.

[GOL 89] GOLUB G., VAN LOAN C., *Matrix Computations*, Johns Hopkins University Press, Baltimore, MD, 1989.

[JOH 82] JOHNSON D., "The application of spectral analysis methods to bearing estimation problems", *Proceedings of the IEEE,* vol. 70, p. 1018-1028, 1982.

[KAR 84] KARHUNEN J., "Adaptive algorithms for estimating eigenvectors of correlation type matrices", *Proc. IEEE ICASSP,* San Diego, CA, p. 14.6.1-14.6.4., March 1984.

[KAR 86] KARASALO I., "Estimating the covariance matrix by signal subspace averaging", *IEEE Trans. Acoust., Speech, Signal Processing,* vol. ASSP-34, p. 8-12, 1986.

[KUM 83] KUMARESAN R., TUFTS D., "Estimating the angles of arrival of multiple plane waves", *IEEE Trans. Aerosp. Electron. Syst.,* vol. AES-19, p. 134-139, January 1983.

[LEC 89] LE CADRE J., "Parametric methods for spatial signal processing in the presence of unknown colored noise fields", *IEEE Trans, on Acoustics, Speech and Signal Processing,* vol. ASSP-37, p. 965-983, July 1989.

[MAR 87] MARPLE JR. S., *Digital Spectral Analysis with Applications*, Prentice Hall, Signal Processing series, Alan V. Oppenheim, series editor, 1987.

[MAR 90a] MARCOS S., BENIDIR M., "On a high resolution array processing method non-based on the eigenanalysis approach", *Proceedings of ICASSP,* Albuquerque, New Mexico, p. 2955-2958, 1990.

[MAR 90b] MARCOS S., BENIDIR M., "Source bearing estimation and sensor positioning with the Propagator method", *Advanced Signal Processing Algorithms, Architectures and Implementations,* F.T. Luk editor, SPIE-90, San Diego, CA, p. 312-323, 1990.

[MAR 94] MARSAL M., MARCOS S., "A reduced complexity ESPRIT method and its generalization to an antenna of partially unknown shape", *Proc. ICASSP 94,* Adelaide, Australia, p. IV29-IV.32, 1994.

[MAR 95] MARCOS S., MARSAL A., BENIDIR M., "The propagator method for source bearing estimation", *Signal Processing,* vol. 42, no. 2, p. 121-138, 1995.

[MAR 96] MARCOS S., BENIDIR M., "An adaptive tracking algorithm for direction finding and array shape estimation in a nonstationary environment", special number on "Array optimization and adaptive tracking algorithms", in *Journal of VLSI Signal Processing,* vol. 14, p. 107-118, October 1996.

[MAR 97] MARCOS S., SANCHEZ-ARAUJO J., "Méthodes linéaires haute résolution pour l'estimation de directions d'arrivée de sources. Performances asymptotiques et complexité", *Traitement du Signal,* vol. 14, no. 2, 1997.

[MAR 98] MARCOS S., Ed., *Méthodes à haute résolution. Traitement d'antenne et Analyse Spectrale,* Signal Processing collection, Hermès, Paris, 1998.

[MUN 91] MUNIER J., DELISLE G., "Spatial analysis using new properties of the cross-spectral matrix", *Trans, on Signal Processing,* vol. 39, no. 3, p. 746-749, 1991.

[OWS 78] OWSLEY N., "Adaptive data orthogonalization", *Proc. IEEE ICASSP,* p. 109-112, 1978.

[PIL 89] PILLAI S., *Array Signal Processing*, Springer-Verlag, 1989.

[PIS 73] PISARENKO V., "The retrieval of harmonics from a covariance function", *Geophys. J. Royal Astron. Soc,* vol. 33, p. 347-366, 1973.

[RED 82] REDDY V., EGARDT B., KAILATH T., "Least squares type algorithm for adaptive implementation of Pisarenko's harmonic retrieval method", *IEEE Trans Acoust., Speech, Signal Processing,* vol. ASSP-30, p. 241-251, 1982.

[RIS 78] RISSANEN J., "Modelling by shortest data description", *IEE Automatica,* vol. 14, p. 465-471, 1978.

[ROY 86] ROY R., PAULRAJ A., KAILATH T., "ESPRIT – a subspace rotation approach to estimation of parameters of cisoids in noise", *IEEE Trans. Acoust., Speech, Signal Processing,* vol. ASSP-34, p. 1340-1342, 1986.

[SCH 78] SCHWARTZ G., "Estimating the dimension of a model", *IEEE Ann. Sta.,* vol. 6, p. 461-464, 1978.

[SCH 79] SCHMIDT R., "Multiple emitter location and signal parameter estimation", *RADC Spectral Estimation Workshop*, Rome, NY, p. 243-258, 1979.

[SCH 89] SCHREIBER R., "Implementation of adaptive array algorithms", *IEEE Trans. on Acoustics, Speech and Signal Processing,* vol. ASSP-37, no. 5, p. 1550-1556, 1989.

[STO 92] STOICA P., SÖDERSTROM T., SIMONYTE V., "On estimating the noise power in array processing", *Signal Processing,* vol. 26, p. 205-220, 1992.

[YAN 89] YANG X., SARKAR T., ARVAS E., "A survey of conjugate gradient algorithms for solution of extreme eigen-problem of a symmetric matrix", *IEEE Trans. on Acoustics, Speech and Signal Processing,* vol. ASSP-37, p. 1550-1556, 1989.

[YAN 95] YANG B., "Projection approximation subspace tracking", *IEEE Trans. Signal Processing,* vol. 43, p. 95-107, 1995.

[YEH 86] YEH C., "Projection approach for bearing estimations", *IEEE Trans. on Acoustics, Speech and Signal Processing,* vol. ASSP-34, no. 5, p. 1347-1349, 1986.

[YEH 87] YEH C., "Simple computation of projection matrix for bearing estimations", *IEEE Part F,* vol. 134, no. 2, p. 146-150, 1987.

[ZHA 89] ZHANG Q., WONG K., YIP P., REILLY J., "Statistical analysis of the performance of information theoretic criteria in the detection of the number of signals in array processing", *IEEE Trans. on Acoustics, Speech and Signal Processing,* vol. ASSP-37, p. 1557-1567, 1989.

Chapter 9

Introduction to Spectral Analysis of Non-Stationary Random Signals

The spectral analysis of non-stationary random signals can be made, *as in the stationary case*, by using two large categories of methods:

– non-parametric methods;

– parametric methods.

In each of these large categories, we also find two types of methods specific to the non-stationary case. Actually, in non-stationary context, the spectral analysis can be made:

– either by using the stationary tools and by adapting them to the non-stationary nature of signals;

– or by introducing new techniques.

Thus, the reader will find three parts in this chapter: a first part which makes it possible to set out the problem of defining the "evolutive spectrum", a second part which tackles the non-parametric techniques and the third part which creates a panorama of the parametric methods used in a non-stationary context.

A more complete panorama of the tools for spectral analysis of non-stationary random signals can be found in [HLA 05].

Chapter written by Corinne MAILHES and Francis CASTANIÉ.

9.1. Evolutive spectra

9.1.1. *Definition of the "evolutive spectrum"*

The problem of the spectral representation of a random process is a difficult problem which deserves our attention. Loynes [LOY 68] formulates a set of properties, which should be those of any spectral representation, all these properties being verified in the stationary case. These properties are listed in section 9.1.2. He also reviews five different definitions of spectral representations. Among these, Priestley's [PRI 65, PRI 91] is based on oscillatory random processes. The definition is only valid for a reduced and badly defined class of random processes since linear combination of oscillatory processes is not necessarily oscillatory. Instead of this definition, Grenier [GRE 81a, GRE 81b] prefers that of Melard [MEL 78] and Tjøstheim [TJØ 76], which is built based on the (unique) canonical decomposition of any random process $x(t)$ [LAC 00]:

$$x(t)=\sum_{u=-\infty}^{t} h(t,u)\varepsilon(u)+v(t) \qquad [9.1]$$

$\varepsilon(t)$ represents the process of innovation of $x(t)$ such that $E\left[\varepsilon(t)\varepsilon(u)\right]=\delta(t-u)$ and $v(t)$ corresponds to the process part referred to as "singular".

Starting from this canonical decomposition, Mélard and Tjøstheim propose to define the evolutive spectrum as:

$$s(t,f)=\frac{1}{2\pi}\left|\sum_{u=-\infty}^{t} h(t,u)\exp(j2\pi fu)\right|^2 \qquad [9.2]$$

This definition meets a large number of the properties listed by Loynes.

However, a paradox can be highlighted with this definition of evolutive spectrum [9.2]. Actually, using very precise hypotheses, we can show that identical innovations for $t > p$ (p being a parameter that intervenes in the hypotheses) lead to evolutive spectra which become identical only for $t \to \infty$. Within the framework of parametric modeling in a stationary context, this problem causes Grenier [GRE 81a, GRE 81b] to replace [9.1] by a status space model in an observable canonical form:

$$\mathbf{y}(t)=\mathbf{A}\mathbf{y}(t-1)+\mathbf{b}\varepsilon(t)$$

$$x(t) = (100\cdots0)\mathbf{y}(t)$$

with:

$$A = \begin{pmatrix} -a_1(t) & -a_2(t) & \cdots & -a_n(t) \\ 1 & 0 & \cdots & 0 \\ \vdots & \ddots & \ddots & \vdots \\ 0 & \cdots & 1 & 0 \end{pmatrix} \quad \text{and} \quad \mathbf{b} = \left[b_0(t)\cdots b_p - 1(t)\right]^T$$

Thus, the evolutive spectrum is rational and defined by:

$$s(t,f) \triangleq \frac{1}{2\pi}\left|\frac{\breve{b}(t,z)\breve{b}\left(t,z^{-1}\right)}{\breve{a}(t,z)\breve{a}\left(t,z^{-1}\right)}\right|_{z=e^{j2\pi f}} \qquad [9.3]$$

where $\breve{a}(t,z)$ and $\breve{b}(t,z)$ are written:

$$\breve{a}(t,z) = 1 + a_1(t)z^{-1} + \cdots + a_n(t)z^{-n}$$
$$\breve{b}(t,z) = b_0(t)b_1(t)z^{-1} + \cdots + b_{-n}(t)z^{-n+1}$$

Grenier [GRE 81a, GRE 81b] shows that the rational evolutive spectrum behaves in a more satisfactory way than the evolutive spectrum of equation [9.2].

9.1.2. *Evolutive spectrum properties*

The evolutive spectrum defined this way observes a large number of the properties desired by Loynes [LOY 68]:

– the variance of the process can be written:

$$E\left[x(t)^2\right] = \sigma_x^2(t) = \int_{-\infty}^{+\infty} s(t,f)\,df$$

– *s (t, f)* coincides in the stationary case with the power spectral density of the process;

– the evolutive spectrum is a real and positive function of *t* and *f*;

– if the process is real, $s(t,f) = s(t,-f)$;

– multiplying the process by a complex exponential $\exp(-j2\pi f_0 t)$ means shifting (modulo 1) its evolutive spectrum by f_0: the corresponding evolutive spectrum is $s(t, f + f_0)$;

– if $x_2(t) = x_1(t+h)$ then $s_2(t,f) = s_1(t+h,f)$;

– if the non-stationary process $x(t)$ is identical to the rational process $x_1(t)$ for the negative times and to the purely autoregressive process $x_2(t)$ of order p for the strictly positive times, then:

$$s(t,f) = s_1(t,f) \quad t \leq 0$$
$$s(t,f) = s_2(t,f) \quad t > p$$

when the processes $x_1(t)$ and $x_2(t)$ are non-correlated, whether they are stationary or non-stationary.

9.2. Non-parametric spectral estimation

The major interest of the non-parametric spectral estimation is of not making *a priori* hypothesis on the signal (as the parametric methods do) and of providing a "blind" analysis tool which is applicable to any type of signal.

Among the non-parametric methods used in non-stationary context, the oldest is the sliding Fourier transform (also called "the short-term" Fourier transform). The Fourier transform was adapted with the introduction of a time window of a sufficiently small size so that the analyzed signal is quasi-stationary along this window. The spectrogram that results from this is defined as:

$$S_{TF}(t,f) = |X(t,f)|^2$$

where $X(t, f)$ designates the Fourier transform of the weighted signal by an analysis time window $w(t)$ that we move in time:

$$X(t,f) = \int_{\mathbb{R}} x(u)\, w(u-t) \exp(-j2\pi f u)\, du$$

The major obstacle of this technique is the compulsory compromise to be made between the time resolution and the spectral resolution that we wish to obtain. With the shortening of the size of the analysis window, the time localization improves to the detriment of the spectral localization and vice versa.

For signals with rapid variations, this solution is thus not very satisfactory. Historically, the Wigner-Ville distribution then appeared. This time-frequency tool verifies a large number of desired properties of the time-frequency representations, which explains its success. It is defined in the following way:

$$W(t,f)=\int_{\mathbb{R}} x\left(t+\frac{u}{2}\right)x^*\left(t-\frac{u}{2}\right)\exp(-j2\pi fu)\,du$$

It preserves the localizations in time and in frequency and allows an exact description of the modulations of linear frequencies. On the other hand, it can take negative values and its bilinear structure generates interferences between the different components of a signal, sometimes making its interpretation difficult. In order to reduce the interferential terms, smoothing is introduced. We obtain then the smoothed pseudo Wigner-Ville distribution:

$$W(t,f)=\iint_{\mathbb{R}^2}\left|w_f\left(\frac{u}{2}\right)\right|^2 w_t(v-t)\,x\left(v+\frac{u}{2}\right)x^*\left(v-\frac{u}{2}\right)\exp(-j2\pi fu)\,dvdu$$

We can generalize this time-frequency representation by the formulation of a class known as Cohen's class:

$$C(t,f)=\frac{1}{2\pi}\iiint_{\mathbb{R}^3}\Phi(\theta,\tau)\,x\left(u+\frac{\tau}{2}\right)x^*\left(u-\frac{\tau}{2}\right)\exp\left(-j(\theta t-2\pi\tau f-\theta u)\right)dud\tau d-$$

Φ is an arbitrary kernel. The definition of the kernel is sufficient to determine a representation. The desired properties of a time-frequency representation induce certain conditions on the kernel. Based on this, the Wigner-Ville transform can be derived but we can also introduce the distributions of Choi-Williams, Zao-Atlas-Marx, Page, Rihaczek, etc. Thus, the user finds a set of time-frequency representations at his disposal, each one having its own characteristics and favoring a part of the desirable properties according to the case of interest. For example, if the time localization of a frequency jump is a primordial criterion, the Zao-Atlas-Marx distribution is without any doubt the most interesting to use.

Further details on these time-frequency representations can be found in [HLA 05] and [FLA 93].

9.3. Parametric spectral estimation

Applying parametric models in a non-stationary context is interesting in so far as that makes it possible to take advantage of the good frequency resolution of these methods on relatively short windows and to have numerous noise reduction algorithms available.

The use of a parametric method puts the problem of *a priori* knowledge on the signal because the model should be chosen so that it is the closest possible to the kind of signal type studied. In a non-stationary context, the parametric modeling is used by modifying the different kinds of models known in the stationary case in order to adapt them to the context. A complete chapter of [HLA 05] is dedicated to the use of the parametric models in the case of non-stationary signals.

There are chiefly two ways of using the parametric modeling methods in a non-stationary context. The first way consists of supposing a local stationarity and of using a method of stationary parametric modeling on short observation windows. The second way consists of using these parametric models to modify these models by suppressing a stationarity condition and by making them thus adequate for modeling a non-stationary phenomenon.

9.3.1. *Local stationary postulate*

We can discriminate two categories among the parametric modeling methods postulating the local stationarity: first, the *sliding* methods, and second, the *adaptive* and *recursive* methods.

The *sliding* methods use the stationary parametric models on a short observation window which is shifted step-by-step along the time axis. The model parameters noted θ (*k*) are thus estimated at time *k* according to an *a priori* observation window of fixed size, which is progressively shifted along the time axis:

$$\hat{\theta}(k)=\hat{\theta}\left(\left[x(k),x(k-1),\ldots,x(k-N)\right]\right)$$

The time increment corresponds to the step of the window displacement and which is chosen by the user. It can be of a sample yielding a high computational cost for an interest, which sometimes can be minimal. If the signal evolution is slow, there is no need to shift this observation window (which corresponds also to the parameter estimation window) with a one sample increment. Rather, an M sample increment should be used. The successive spectral analyses will be thus made on every *M* sample. The size of the analysis window is an important parameter of this kind of method. On one hand, it should be short enough to make it possible to postulate the local stationarity, but on the other hand, it conditions the maximum order of the model and thus of the spectral resolution.

In the *adaptive and recursive* methods, the parameters are enriched in a recursive way over time, which often gives them the property to "adapt" themselves to a non-stationary context. The modeling is called adaptive if its parameters are modified using a given criterion as soon as a new signal value is known. These techniques are often used for real time applications. Essentially, there are two types of stationary modeling that are made adaptive: first, state space model the autoregressive moving average (ARMA) modeling *via* the state space model and Kalman modeling, with the particular case of adaptive AR [FAR 80] and second, Prony modeling [CAS 88, LAM 88]. The framework of the adaptive ARMA is based more particularly in the framework on a stochastic approach by using its state space model and the Kalman filtering [NAJ 88]; these aspects are not tackled in this book.

In a context of non-stationary parametric modeling, the classic adaptive methods have memories with exponential forgeting strategies and can be regrouped in two families of algorithms: the gradient methods and the least-squares methods. For more information on adaptive algorithms, see [BEL 89, HAY 90].

9.3.2. *Elimination of a stationary condition*

In order to build a usable parametric model in a non-stationary context, we can modify the parametric models known and used in a stationary context by eliminating one of the stationarity condition. The model then becomes intrinsically non-stationary and can be applied in the framework of a spectral analysis.

The unstable models and the models with variable or evolutive parameters can be mentioned amongst all the models built this way.

In unstable models, a part of the poles is on the unit circle, making them unstable and likely to model certain classes of particular non-stationarities. The ARIMA models and the seasonal models [BOX 70] are amongst the most well known. Box and Jenkins [BOX 70] have largely contributed to the popularization of the ARIMA models (p, d, q): AutoRegressive Integrated Moving Average. This is a category of non-stationary models which make it possible to describe only one type of non-stationarity of the "polynomial" type. The ARIMA model can be seen as a particular case of ARMA model ($p + d$, q) in which the polynomial of the autoregressive part is of the form:

$$\breve{a}(z) = \left(1 - z^{-1}\right)^d \breve{a}_p(z)$$

with d an integer and:

$$\breve{a}_p(z)=1+\sum_{n=1}^{p}a_n z^{-n} \tag{9.4}$$

a polynomial whose roots are inside the unit circle.

The models with variable or evolutive parameters correspond to the parametric models in which the parameters vary in time. The evolutive models that are most well known and used are the evolutive ARMA and more particularly the evolutive AR. Actually, the ARMA models can be used for the modeling of non-stationary signals by supposing that the parameters of the model progress over time. The first problem is to define the evolution of the coefficients of the model. A retained solution is the projection of these coefficients on a finite basis of functions: the parameters of the model are expressed in the form of linear combinations of functions $f_m(k)$ of a predefined basis. This class of models was introduced by [LIP 75, MEN 91, RAO 70] and popularized by [GRE 86]. The process $x(k)$ is thus modeled as being the output of a recursive linear filter with coefficients that vary in time, excited by an entry of the Gaussian white noise type $n(k)$:

$$x(k)=\sum_{l=1}^{p}a_l(k-l)x(k-l)+\sum_{l=0}^{q}b_l(k-l)n(k-l) \tag{9.5}$$

with

$$a_l(k)=\sum_{m=1}^{M}a_{lm}f_m(k)$$

$$b_l(k)=\sum_{m=1}^{M}b_{lm}f_m(k)$$

In order to make the writing easier, the degree of the basis M is chosen as identical for the AR part and for the MA part, but in practice it will be natural to consider different degrees.

The evolutive AR model is a particular case of the evolutive ARMA model in which the MA part is reduced to an order 0.

The choice of the basis of functions is an important step of the modeling process. Choosing an orthogonal basis is not necessary for the identification of the model but improves the estimation by a better numerical conditioning of the systems to solve.

According to the non-stationarity type, the optimal base is that which will make it possible to represent in the best possible way (lowest possibility of approximation error) the AR and MA coefficients with a minimum number of base functions. The bases of the functions usually used are the time powers, the Legendre basis and the Fourier basis.

The concept of evolutive parameters is also found in the definition of the Kamen evolutive poles [KAM 88] or through the evolutive Prony model [MOL 95], extensions of the stationary Prony model in which the poles have a trajectory which varies over time.

For the models with non-stationary inputs, among which the multi-pulse model is without any doubt the most popular, the multi-pulse modeling was proposed for the speech coding by Atal and Remde [ATA 82]. This consists of determining the impulse input (series of excitation pulses) which, by feeding a linear translation invariant system (of impulse response $h(k)$), will make it possible to describe the signal $x(k)$ by:

$$x(k) = \sum_{m=1}^{L} a_m h(k - k_m) + e(k)$$

The epochs of the excitation pulses $\{k_m\}$ are not necessarily regularly distributed. Thus, starting from a determined waveform $h(k)$, either by *a priori* knowledge or by a preliminary modeling (AR modeling in general), we search the indicating pulses $\{k_m, a_m\}$ which minimize a quadratic error criterion.

This type of approach makes it possible to model signals presenting transients, echoes or multi-paths. In addition to speech coding (under Atal "pulse" [ATA 82]), this type of modeling was used in seismics [COO 90, GUE 86, HAM 88], in electromyography [GAS 88], etc. A summary of multi-pulse modeling as well as a thorough analysis of the method and of its different variations can be found in [GAS 93]. In particular, there we will find models associating the Prony modeling and the multi-pulse. The "multimodel multi-pulse" approach and the link existing between AR modeling and Prony modeling leads to the visualization of the generalization of the multi-pulse modeling by systematically associating a Prony modeling to each waveform. Thus, the analysis of multi-pulse Prony means considering the following model:

$$\hat{x}(k) = \sum_{n=1}^{L} \sum_{m=1}^{P_k} B_{mn} z_{mn}^{k-k_n} U(k - k_n)$$

in which $U(k)$ designates the heavyside function.

9.3.3. *Application to spectral analysis*

The sliding and adaptive methods make it possible to estimate a vector parameter according to time. The time-frequency analysis naturally results from this estimation by defining an evolutive spectrum identical to the spectral estimator of the corresponding stationary model in which we replace the stationary parameters by the estimated time-varying parameters.

Thus, the AR sliding or AR adaptive modeling leads to the estimation of the vector of the time-varying autoregressive coefficients $a(t)$. The time-frequency analysis resulting from the sliding or adaptive AR modeling consists of plotting the following spectral estimator in the time-frequency plane:

$$s(t,f)=\frac{1}{2\pi}\frac{\sigma_{\varepsilon}^{2}(t)}{\left|\breve{a}(t,z)\breve{a}\left(t,z^{-1}\right)\right|_{z=e^{j2\pi f}}}$$

$\sigma_{\varepsilon}^{2}(t)$ representing the power of the excitatory white noise at each moment (or at each adaptation step).

We will note the similarity of this AR sliding or adaptive spectrum with the definition of the evolutive spectrum of equation [9.3].

Also, the sliding or adaptive Prony modeling leads to estimating time-varying complex amplitudes $\{b_n(t)\}_{n=1,p}$ and poles $\{z_n(t)\}_{n=1,p}$. This type of modeling makes it possible to plot a spectral estimator in the time-frequency plane; this estimator is defined by:

$$s(t,f)=\left|\sum_{n=1}^{P} b_n(t)\frac{1-z_n^2(t)}{\left(1-z_n(t)z^{-1}\right)\left(1-z_n(t)z\right)}\right|_{z=e^{j2\pi f}}$$

It is also possible to use these sliding or adaptive methods in a different way. In the context of mode tracking, the final product of the time-frequency analysis can be the plot of the modes (i.e. the poles) estimated over time. Actually, the AR sliding or adaptive modeling makes it possible to estimate the poles by solving at each moment of:

$$\breve{a}(t,z)=0$$

The sliding or adaptive Prony modeling directly provides the estimation of these time-varying poles according to time.

By selecting only the poles whose module is close enough to the unit circle, we can thus follow the evolution in time of the frequencies characterizing the signal under interest, as well as their amplitudes in the case of Prony modeling.

The unstable models, the evolutive models and the multi-pulse models make up the class of non-stationary parametric models built starting from stationary models by eliminating one or several stationarity conditions. These models can have multiple applications, among which the time-frequency analysis has an important place.

For the unstable models or evolutive ARMA, the expression of a transmittance in z leads to the definition of an evolutive spectrum according to equation [9.3], which allows a time-frequency analysis of the process.

The multi-pulse methods authorize a time-frequency analysis particularly well adapted to the case of non-stationary signals made up of a sum of modes appearing at different moments. In a time-frequency analysis context, this type of modeling allows the definition of a rational spectrum for the case of determinist signals.

The monomodel case leads to:

$$\breve{x}(z) = \frac{\sum_{m=1}^{L} a_m z^{-k_m}}{\sum_{m=1}^{P} b_m z^{-m}}$$

where the $\{b_m\}_{m=1,p}$ correspond to the coefficients of the AR model used for modeling the waveform $h(k)$. We thus get to the expression of an ARMA rational spectrum of a very high degree in MA (of the order of the length N of the observation window).

9.4. Bibliography

[ATA 82] ATAL B. S., REMDE J. M., "A new model of LPC excitation for producing natural speech at low bit rates", *Proc. IEEE ICASSP-82,* p. 614-618, 1982.

[BEL 89] BELLANGER M., *Analyse des signaux et filtrage numérique adaptatif,* Masson, 1989.

[BOX 70] BOX G., JENKINS G., *Time Series Analysis, Forecasting and Control,* Holden-Day, San Francisco, 1970.

[CAS 88] CASTANIÉ F., LAMBERT-NEBOUT C., "A fast and parallel algorithm to solve the Vandermonde system in Prony modeling", *Proc. of EUSIPCO 88,* Grenoble, p. 571-574, September 1988.

[COO 90] COOKEY M., TRUSSELL H., WON I., "Seismic deconvolution by multi-pulse methods", *IEEE Trans. Acoust., Speech, Signal Processing,* vol. 38, no. 1, p. 156-159, January 1990.

[FAR 80] FARGETTON H., GENDRIN R., J.L. L., "Adaptive methods for spectral analysis of time varying signals", *Proc. of EUSIPCO 80,* Lausanne, Switzerland, p. 777-792, September 1980.

[FLA 93] FLANDRIN P., *Temps-fréquence,* Signal processing series, Hermès, 1993.

[GAS 88] GASMI F., SITBON S., CASTANIÉ F., "Multi-pulse excitation modeling applied to EMG signals", *Proc. EUSIPCO 88,* p. 1271-1274 May 1988.

[GAS 93] GASMI F., La modélisation multi-impulsionnelle, PhD Thesis, INP Toulouse, 1993.

[GRE 81a] GRENIER Y., "Estimation de spectres rationnels non-stationnaires", *Proc. of GRETSI 1981,* Nice, France, p. 185-192, 1981.

[GRE 81b] GRENIER Y., "Rational non-stationary spectra and their estimation", *1st ASSP Workshop on Spectral Estimation,* Hamilton, Ontario, p. 6.8.1-6.8.8., 1981.

[GRE 86] GRENIER Y., "Modèles ARMA à coefficients dépendant du temps: estimateurs et applications", *Traitement du Signal,* vol. 3, no. 4, p. 219-233, 1986.

[GUE 86] GUEGUEN C., MOREAU N., "Determining MA models as salvos of pulses", *Proc. of ICASSP 86,* p. 12.4.1-12.4.4., 1986.

[HAM 88] HAMIDI R., BOUCHER J., J.L. B., "Combined used of homomorphic filtering and multi-pulse modeling for the deconvolution of seismic reflection signals", *Proc. of EUSIPCO 88,* p. 975-978, 1988.

[HAY 90] HAYKIN S., *Adaptive Filter Theory,* Prentice Hall, 1990.

[HLA 05] HLAWATSCH R., AUGER R., OVARLEZ J.-P., *Temps-fréquence: concepts et outils,* IC2 series, Hermès Sciences, Paris, 2005.

[KAM 88] KAMEN E., "The poles and zeros of a linear time-varying system", *Linear Algebra and its Application,* vol. 98, p. 263-289, 1988.

[LAC 00] LACAZE B., *Processus aléatoires pour les communications numériques,* Hermès, Paris, 2000.

[LAM 88] LAMBERT-NEBOUT C., CASTANIÉ R., "An adaptive Prony algorithm", *Proc. of EUSIPCO 88,* p. 403-406, September 1988.

[LIP 75] LIPORACE L., "Linear estimation of non-stationary signals", *Journal of Acoust. Soc. Amer,* vol. 58, no. 6, p. 1288-1295, 1975.

[LOY 68] LOYNES R., "On the concept of the spectrum for non-stationary processes", *Journal of the Royal Statist. Soc,* Series B, vol. 30, no. 1, p. 1-30, 1968.

[MEL 78] MELARD G., "Propriétés du spectre évolutif d'un processus non-stationnaire", *Ann. Inst. H. Poincaré*, Section B, vol. 14, no. 4, p. 411-424, 1978.

[MEN 91] MENDEL J., "Tutorial on high-order statistics in signal processing and system theory: theoretical results and some applications", *Proceedings of the IEEE,* vol. 79, no. 3, p. 277-305, 1991.

[MOL 95] MOLINARO R., CASTANIE R., "Modèle de Prony à pôles dépendant du temps", *Traitement du Signal,* vol. 12, no. 5, p. 421-431, 1995.

[NAJ 88] NAJIM M., *Modélisation et identification en traitement du signal,* Masson, 1988.

[PRI 65] PRIESTLEY M., "Evolutionary spectra and non-stationary processes", *Journal of the Royal Statist. Soc,* Series B, vol. 27, no. 2, p. 204-237, 1965.

[PRI 91] PRIESTLEY M., *Non-linear and Non-stationary Time Series Analysis*, Academic Press, 1991.

[RAO 70] RAO T., "The fitting of non-stationary time-series models with time-dependent parameters", *J. of the Royal Statist. Soc,* Series B, vol. 32, no. 2, p. 312-322, 1970.

[TJØ 76] TJØSTHEIM D., "Spectral generating operators for non-stationary processes", *Adv. Appl. Prob.,* vol. 8, p. 831-846, 1976.

List of Authors

Olivier BESSON
Département avionique et systèmes
Ecole nationale supérieure d'ingénieurs de constructions aéronautiques
Toulouse

Francis CASTANIÉ
IRIT-TESA
Ecole nationale supérieure d'électrotechnique, d'électronique, d'informatiques, d'hydraulique et des télécommunications
Toulouse

André FERRARI
LUAN
University of Nice Sophia-Antiopolis

Eric LE CARPENTIER
IRCCyN
Ecole centrale de Nantes

Corinne MAILHES
IRIT-TESA
Ecole nationale supérieure d'électrotechnique, d'électronique, d'informatiques, d'hydraulique et des télécommunications
Toulouse

Sylvie MARCOS
LSS
CNRS
Gif-sur-Yvette

Nadine MARTIN
LIS
CNRS
Grenoble

Index

H, I

K, L

M

N

P

R, S

T, W